Using **R** and
RStudio
for Data Management,
Statistical Analysis,
and Graphics

Second Edition

Using R and

RStudio

for Data Management, Statistical Analysis, and Graphics

Second Edition

Nicholas J. Horton

Department of Mathematics and Statistics
Amherst College
Massachusetts, U.S.A.

Ken Kleinman

Department of Population Medicine
Harvard Medical School and
Harvard Pilgrim Health Care Institute
Boston, Massachusetts, U.S.A.

CRC Press
Taylor & Francis Group
Boca Raton   London   New York

CRC Press is an imprint of the
Taylor & Francis Group, an **informa** business

A CHAPMAN & HALL BOOK

CRC Press
Taylor & Francis Group
6000 Broken Sound Parkway NW, Suite 300
Boca Raton, FL 33487-2742

© 2015 by Taylor & Francis Group, LLC
CRC Press is an imprint of Taylor & Francis Group, an Informa business

No claim to original U.S. Government works

International Standard Book Number-13: 978-1-4822-3736-8 (Hardback)

Visit the Taylor & Francis Web site at
http://www.taylorandfrancis.com

and the CRC Press Web site at
http://www.crcpress.com

Contents

List of Tables

List of Figures

Preface to the second edition

Software systems such as R evolve rapidly, and so do the approaches and expertise of statistical analysts.

In 2009, we began a blog in which we explored many new case studies and applications, ranging from generating a Fibonacci series to fitting finite mixture models with concomitant variables. We also discussed some additions to R, the RStudio integrated development environment, and new or improved R packages. The blog now has hundreds of entries and according to Google Analytics has received hundreds of thousands of visits.

The volume you are holding is a larger format and longer than the first edition, and much of the new material is adapted from these blog entries, while it also includes other improvements and additions that have emerged in the last few years.

We have extensively reorganized the material in the book and created three new chapters. The firsts, "Simulation," includes examples where data are generated from complex models such as mixed-effects models and survival models, and from distributions using the Metropolis–Hastings algorithm. We also explore interesting statistics and probability examples via simulation. The second is "Special topics," where we describe some key features, such as processing by group, and detail several important areas of statistics, including Bayesian methods, propensity scores, and bootstrapping. The last is "Case studies," where we demonstrate examples of useful data management tasks, read complex files, make and annotate maps, show how to "scrape" data from the web, mine text files, and generate dynamic graphics.

We also describe RStudio in detail. This powerful and easy-to-use front end adds innumerable features to R. In our experience, it dramatically increases the productivity of R users, and by tightly integrating reproducible analysis tools, helps avoid error-prone "cut and paste" workflows. Our students and colleagues find RStudio an extremely comfortable interface.

We used a reproducible analysis system (`knitr`) to generate the example code and output in the book. Code extracted from these files is provided on the book website. In this edition, we provide a detailed discussion of the philosophy and use of these systems. In particular, we feel that the `knitr` and `markdown` packages for R, which are tightly integrated with RStudio, should become a part of every R user's toolbox. We can't imagine working on a project without them.

The second edition of the book features extensive use of a number of new packages that extend the functionality of the system. These include `dplyr` (tools for working with dataframe-like objects and databases), `ggplot2` (implementation of the Grammar of Graphics), `ggmap` (spatial mapping using `ggplot2`), `ggvis` (to build interactive graphical displays), `httr` (tools for working with URLs and HTTP), `lubridate` (date and time manipulations), `markdown` (for simplified reproducible analysis), `shiny` (to build interactive web applications), `swirl` (for learning R, in R), `tidyr` (for data manipulation), and `xtable` (to create publication-quality tables). Overall, these packages facilitate ever more sophisticated analyses.

Finally, we've reorganized much of the material from the first edition into smaller, more focused chapters. Readers will now find separate (and enhanced) chapters on data input and output, data management, statistical and mathematical functions, and programming, rather than a single chapter on "data management." Graphics are now discussed in two chapters: one on high-level types of plots, such as scatterplots and histograms, and another on customizing the fine details of the plots, such as the number of tick marks and the color of plot symbols.

We're immensely gratified by the positive response the first edition elicited, and hope the current volume will be even more useful to you.

On the web

The book website at http://www.amherst.edu/~nhorton/r2 includes the table of contents, the indices, the HELP dataset in various formats, example code, a pointer to the blog, and a list of errata.

Acknowledgments

In addition to those acknowledged in the first edition, we would like to thank J.J. Allaire and the RStudio developers, Danny Kaplan, Deborah Nolan, Daniel Parel, Randall Pruim, Romain Francois, and Hadley Wickham, plus the many individuals who have created and shared R packages. Their contributions to R and RStudio, programming efforts, comments, and guidance and/or helpful suggestions on drafts of the revision have been extremely helpful. Above all, we greatly appreciate Sara and Julia as well as Abby, Alana, Kinari, and Sam, for their patience and support.

Amherst, MA
October 2014

Preface to the first edition

R (R development core team, 2009) is a general purpose statistical software package used in many fields of research. It is licensed for free, as open-source software. The system is developed by a large group of people, almost all volunteers. It has a large and growing user and developer base. Methodologists often release applications for general use in R shortly after they have been introduced into the literature. While professional customer support is not provided, there are many resources to help support users.

We have written this book as a reference text for users of R. Our primary goal is to provide users with an easy way to learn how to perform an analytic task in this system, without having to navigate through the extensive, idiosyncratic, and sometimes unwieldy documentation or to sort through the huge number of add-on packages. We include many common tasks, including data management, descriptive summaries, inferential procedures, regression analysis, multivariate methods, and the creation of graphics. We also show some more complex applications. In toto, we hope that the text will facilitate more efficient use of this powerful system.

We do not attempt to exhaustively detail all possible ways available to accomplish a given task in each system. Neither do we claim to provide the most elegant solution. We have tried to provide a simple approach that is easy to understand for a new user, and have supplied several solutions when it seems likely to be helpful.

Who should use this book

Those with an understanding of statistics at the level of multiple-regression analysis should find this book helpful. This group includes professional analysts who use statistical packages almost every day as well as statisticians, epidemiologists, economists, engineers, physicians, sociologists, and others engaged in research or data analysis. We anticipate that this tool will be particularly useful for sophisticated users, those with years of experience in only one system, who need or want to use the other system. However, intermediate-level analysts should reap the same benefit. In addition, the book will bolster the analytic abilities of a relatively new user, by providing a concise reference manual and annotated examples.

Using the book

The book has two indices, in addition to the comprehensive table of contents. These include: 1) a detailed topic (subject) index in English; 2) an R command index, describing R syntax.

Extensive example analyses of data from a clinical trial are presented; see Table B.1 (p. 237) for a comprehensive list. These employ a single dataset (from the HELP study), described in Appendix B. Readers are encouraged to download the dataset and code from the book website. The examples demonstrate the code in action and facilitate exploration by the reader.

In addition to the HELP examples, a case studies and extended examples chapter utilizes many of the functions, idioms and code samples introduced earlier. These include explications of analytic and empirical power calculations, missing data methods, propensity score analysis, sophisticated data manipulation, data gleaning from websites, map making, simulation studies, and optimization. Entries from earlier chapters are cross-referenced to help guide the reader.

Where to begin

We do not anticipate that the book will be read cover to cover. Instead, we hope that the extensive indexing, cross-referencing, and worked examples will make it possible for readers to directly find and then implement what they need. A new user should begin by reading the first chapter, which includes a sample session and overview of the system. Experienced users may find the case studies to be valuable as a source of ideas on problem solving in R.

Acknowledgments

We would like to thank Rob Calver, Kari Budyk, Shashi Kumar, and Sarah Morris for their support and guidance at Informa CRC/Chapman and Hall. We also thank Ben Cowling, Stephanie Greenlaw, Tanya Hakim, Albyn Jones, Michael Lavine, Pamela Matheson, Elizabeth Stuart, Rebbecca Wilson, and Andrew Zieffler for comments, guidance and/or helpful suggestions on drafts of the manuscript.

Above all we greatly appreciate Julia and Sara as well as Abby, Alana, Kinari, and Sam, for their patience and support.

Northampton, MA and Amherst, MA
February, 2010

Chapter 1

Data input and output

This chapter reviews data input and output, including reading and writing files in spreadsheet, ASCII file, native, and foreign formats.

1.1 Input

R provides comprehensive support for data input and output. In this section we address aspects of these tasks. Datasets are organized in dataframes (A.4.6), or connected series of rectangular arrays, which can be saved as platform-independent objects. UNIX-style directory delimiters (forward slash) are allowed on Windows.

1.1.1 Native dataset

Example: 7.10

```
load(file="dir_location/savedfile")    # works on all OS including Windows
or
load(file="dir_location\\savedfile")   # Windows only
```

Note: Forward slash is supported as a directory delimiter on all operating systems; a double backslash is supported under Windows. The file `savedfile` is created by `save()` (see 1.2.3). Running the command `print(load(file="dir_location/savedfile"))` will display the objects that are added to the workspace.

1.1.2 Fixed format text files

See 1.1.9 (read more complex fixed files) and 12.2 (read variable format files).

```
ds = read.table("dir_location\\file.txt", header=TRUE) # Windows only
or
ds = read.table("dir_location/file.txt", header=TRUE)  # all OS (including
                                                       # Windows)
```

Note: Forward slash is supported as a directory delimiter on all operating systems; a double backslash is supported under Windows. If the first row of the file includes the name of the variables, these entries will be used to create appropriate names (reserved characters such as '$' or '[' are changed to '.') for each of the columns in the dataset. If the first row doesn't include the names, the `header` option can be left off (or set to `FALSE`), and the variables

1

will be called `V1`, `V2`, ... `Vn`. A limit on the number of lines to be read can be specified through the `nrows` option. The `read.table()` function can support reading from a URL as a filename (see 1.1.12) or browse files interactively using `read.table(file.choose())` (see 4.3.7).

1.1.3   Other fixed files

See 1.1.9 (read more complex fixed files) and 12.2 (read variable format files)

Sometimes data arrives in files that are very irregular in shape. For example, there may be a variable number of fields per line, or some data in the line may describe the remainder of the line. In such cases, a useful generic approach is to read each line into a single character variable, then use character variable functions (see 2.2) to extract the contents.

```
ds = readLines("file.txt")
```

or

```
ds = scan("file.txt")
```

Note: The `readLines()` function returns a character vector with length equal to the number of lines read (see `file()`). A limit on the number of lines to be read can be specified through the `nrows` option. The `scan()` function returns a vector, with entries separated by white space by default. These functions read by default from standard input (see `stdin()` and `?connections`), but can also read from a file or URL (see 1.1.12). The `read.fwf()` function may also be useful for reading fixed-width files.

1.1.4   Comma-separated value (CSV) files

Example: 2.6.1

```
ds = read.csv("dir_location/file.csv")
```

Note: The `stringsAsFactors` option can be set to prevent automatic creation of factors for categorical variables. A limit on the number of lines to be read can be specified through the `nrows` option. The command `read.csv(file.choose())` can be used to browse files interactively (see 4.3.7). The comma-separated file can be given as a URL (see 1.1.12). The `colClasses` option can be used to speed up reading large files. Caution is needed when reading date and time variables (see 2.4).

1.1.5   Read sheets from an Excel file

```
library(gdata)
ds = read.xls("http://www.amherst.edu/~nhorton/r2/datasets/help.xlsx",
    sheet=1)
```

Note: The sheet number can be provided as a number or a name.

1.1.6   Read data from R into SAS

The R package `foreign` includes the `write.dbf()` function; we recommend this as a reliable format for extracting data from R into a SAS-ready file, though other options are possible. Then SAS `proc import` can easily read the DBF file.

```
tosas = data.frame(ds)
library(foreign)
write.dbf(tosas, "dir_location/tosas.dbf")
```

This can be read into SAS using the following commands:

```
proc import datafile="dir_location\tosas.dbf"
   out=fromr dbms=dbf;
run;
```

1.1.7 Read data from SAS into R

```
library(foreign)
ds = read.dbf("dir_location/to_r.dbf")
```
or
```
library(sas7bdat)
helpfromSAS = read.sas7bdat("dir_location/help.sas7bdat")
```

Note: The first set of code assumes SAS has been used to write out a dataset in DBF format. The second can be used with any SAS formatted dataset; it is based on a reverse-engineering of the SAS dataset format, which SAS has not made public.

1.1.8 Reading datasets in other formats

Example: 6.6.1

```
library(foreign)
ds = read.dbf("filename.dbf")          # DBase
ds = read.epiinfo("filename.epiinfo")  # Epi Info
ds = read.mtp("filename.mtp")          # Minitab portable worksheet
ds = read.octave("filename.octave")    # Octave
ds = read.ssd("filename.ssd")          # SAS version 6
ds = read.xport("filename.xport")      # SAS XPORT file
ds = read.spss("filename.sav")         # SPSS
ds = read.dta("filename.dta")          # Stata
ds = read.systat("filename.sys")       # Systat
```

Note: The `foreign` package can read Stata, Epi Info, Minitab, Octave, SPSS, and Systat files (with the caveat that SAS files may be platform dependent). The `read.ssd()` function will only work if SAS is installed on the local machine.

1.1.9 Reading more complex text files

See 1.1.2 (read fixed files) and 12.2 (read variable format files).

Text data files often contain data in special formats. One common example is date variables. As an example below we consider the following data.

```
1 AGKE 08/03/1999 $10.49
2 SBKE 12/18/2002 $11.00
3 SEKK 10/23/1995 $5.00
```

```
tmpds = read.table("file_location/filename.dat")
id = tmpds$V1
initials = tmpds$V2
datevar = as.Date(as.character(tmpds$V3), "%m/%d/%Y")
cost = as.numeric(substr(tmpds$V4, 2, 100))
ds = data.frame(id, initials, datevar, cost)
rm(tmpds, id, initials, datevar, cost)
```

or (for the date)

```
library(lubridate)
library(dplyr)
tmpds = mutate(tmpds, datevar = mdy(V3))
```

Note: This task is accomplished by first reading the dataset (with default names from `read.table()` denoted `V1` through `V4`). These objects can be manipulated using `as.character()` to undo the default coding as factor variables, and coerced to the appropriate data types. For the `cost` variable, the dollar signs are removed using the `substr()` function. Finally, the individual variables are bundled together as a dataframe. The `lubridate` package includes functions to make handling date and time values easier; the `mdy()` function is one of these.

1.1.10   Reading data with a variable number of words in a field

Reading data in a complex data format will generally require a tailored approach. Here we give a relatively simple example and outline the key tools useful for reading in data in complex formats. Suppose we have data as follows:

```
1 Las Vegas, NV --- 53.3 --- --- 1
2 Sacramento, CA --- 42.3 --- --- 2
3 Miami, FL --- 41.8 --- --- 3
4 Tucson, AZ --- 41.7 --- --- 4
5 Cleveland, OH --- 38.3 --- --- 5
6 Cincinnati, OH 15 36.4 --- --- 6
7 Colorado Springs, CO --- 36.1 --- --- 7
8 Memphis, TN --- 35.3 --- --- 8
8 New Orleans, LA --- 35.3 --- --- 8
10 Mesa, AZ --- 34.7 --- --- 10
11 Baltimore, MD --- 33.2 --- --- 11
12 Philadelphia, PA --- 31.7 --- --- 12
13 Salt Lake City, UT --- 31.9 17 --- 13
```

The `---` means that the value is missing. Note two complexities here. First, fields are delimited by both spaces and commas, where the latter separates the city from the state. Second, cities may have names consisting of more than one word.

```
readcities = function(thisline) {
   thislen = length(thisline)
   id = as.numeric(thisline[1])
   v1 = as.numeric(thisline[thislen-4])
   v2 = as.numeric(thisline[thislen-3])
   v3 = as.numeric(thisline[thislen-2])
   v4 = as.numeric(thisline[thislen-1])
   v5 = as.numeric(thisline[thislen])
   city = paste(thisline[2:(thislen-5)], collapse=" ")
   return(list(id=id,city=city,v1=v1,v2=v2,v3=v3,v4=v4,v5=v5))
}
file =
  readLines("http://www.amherst.edu/~nhorton/r2/datasets/cities.txt")
split = strsplit(file, " ")   # split up fields for each line
as.data.frame(t(sapply(split, readcities)))
```

Note: We first write a function that processes a line and converts each field other than the city name into a numeric variable. The function works backward from the end of the line to find the appropriate elements, then calculates what is left over to store in the city variable. We need each line to be converted into a character vector containing each "word" (character strings divided by spaces) as a separate element. We'll do this by first reading each line, then splitting it into words. This results in a list object, where the items in the list are the vectors of words. Then we call the `readcities()` function for each vector using an invocation of `sapply()` (A.5.2), which avoids use of a `for` loop. The resulting object is transposed, then coerced into a dataframe (see also `count.fields()`).

1.1.11   Read a file byte by byte

It may be necessary to read data that is not stored in ASCII (or other text) format. At such times, it may be useful to read the raw bytes stored in the file.

```
finfo = file.info("full_filename")
toread = file("full_filename", "rb")
alldata = readBin(toread, integer(), size=1, n=finfo$size, endian="little")
```

Note: The `readBin()` function is used to read the file, after some initial prep work. The function requires we input the number of data elements to read. An overestimate is OK, but we can easily find the exact length of the file using the `file.info()` function; the resulting object has a size constituent with the number of bytes. We'll also need a *connection* to the file, which is established in a call to the `file()` function. The `size` option gives the length of the elements, in bytes, and the `endian` option helps describe how the bytes should be read. The `showNonASCII()` and `showNonASCIIfile()` functions can be useful to find non-ASCII characters in a vector or file, respectively.

1.1.12   Access data from a URL

Examples: 5.7.1, 12.4.2, and 12.4.6

```
ds = read.csv("http://www.amherst.edu/~nhorton/r2/datasets/help.csv")
```

or

```
library(RCurl)
myurl = getURL("https://example.com/file.txt")
ds = readLines(textConnection(myurl))
```

Note: The `read.csv()` function, like others that read files from outside R, can access data from a URL. The `readLines()` function reads arbitrary text. To read https (Hypertext Transfer Protocol Secure) URLs, the `getURL()` function from the `RCurl` package is needed, in conjunction with the `textConnection()` function (see also `url()`). Access through proxy servers as well as specification of username and passwords is provided by the function `download.file()`. The `source_DropboxData()` function in the `repmis` package can facilitate reading data from `Dropbox.com`.

1.1.13   Read an XML-formatted file

A sample (flat) XML form of the HELP dataset can be found at `http://www.amherst.edu/~nhorton/r2/datasets/help.xml`. The first ten lines of the file consist of:

```
<?xml version="1.0" encoding="iso-8859-1" ?>
<TABLE>
   <HELP>
      <id> 1 </id>
      <e2b1 Missing="." />
      <g1b1> 0 </g1b1>
      <i11 Missing="." />
      <pcs1> 54.2258263 </pcs1>
      <mcs1> 52.2347984 </mcs1>
      <cesd1> 7 </cesd1>
```

Here we consider reading simple files of this form. While support is available for reading more complex types of XML files, these typically require considerable additional sophistication.

```
library(XML)
urlstring = "http://www.amherst.edu/~nhorton/r2/datasets/help.xml"
doc = xmlRoot(xmlTreeParse(urlstring))
tmp = xmlSApply(doc, function(x) xmlSApply(x, xmlValue))
ds = t(tmp)[,-1]
```

Note: The `XML` package provides support for reading XML files. The `xmlRoot()` function opens a connection to the file, while `xmlSApply()` and `xmlValue()` are called recursively to process the file. The returned object is a character matrix with columns corresponding to observations and rows corresponding to variables, which in this example are then transposed. JSON (JavaScript Object Notation) is a low-overhead alternative to XML. Support for operations using JSON is available in the `RJSONIO` package on Omegahat.

1.1.14   Read an HTML table

Example: 12.4.4

HTML tables are used on websites to arrange data into rows and columns. These can be accessed as objects within R.

```
library(XML)
tables = readHTMLTable(URL)
table1 = result[[1]]
```

Note: In this example, all of the tables in the specified URL are downloaded, and the contents of the first are stored in an object called `table1`.

1.1.15 Manual data entry

```
x = numeric(10)
data.entry(x)
```
or
```
x1 = c(1, 1, 1.4, 123)
x2 = c(2, 3, 2, 4.5)
```

Note: The `data.entry()` function invokes a spreadsheet that can be used to edit or otherwise change a vector or dataframe. In this example, an empty numeric vector of length 10 is created to be populated. The `data.entry()` function differs from the `edit()` function, which leaves the objects given as arguments unchanged, returning a new object with the desired edits (see also the `fix()` function).

1.2 Output

1.2.1 Displaying data

Example: 6.6.2

See 2.1.3 (values of variables in a dataset).

```
dollarcents = function(x)
    return(paste("$", format(round(x*100, 0)/100, nsmall=2), sep=""))
data.frame(x1, dollarcents(x3), xk, x2)
```
or
```
ds[,c("x1", "x3", "xk", "x2")]
```

Note: A function can be defined to format a vector as US dollars and cents by using the `round()` function (see 3.2.4) to control the number of digits (2) to the right of the decimal. Alternatively, named variables from a dataframe can be printed. The `cat()` function can be used to concatenate values and display them on the console (or route them to a file using the `file` option). More control on the appearance of printed values is available through use of `format()` (control of digits and justification), `sprintf()` (use of C-style string formatting) and `prettyNum()` (another routine to format using C-style specifications). The `symnum()` function provides symbolic number coding (this is particularly useful for visualizations of structure matrices).

1.2.2 Number of digits to display

Example: 2.6.1

```
options(digits=n)
```

Note: The `options(digits=n)` command can be used to change the default number of decimal places to display in subsequent R output. To affect the actual significant digits in the data, use the `round()` function (see 3.2.4).

1.2.3   Save a native dataset

Example: 2.6.1

```
save(robject, file="savedfile")
```

Note: An object (typically a dataframe or a list of objects) can be read back into R using `load()` (see 1.1.1).

1.2.4   Creating datasets in text format

```
write.csv(ds, file="full_file_location_and_name")
```
or
```
library(foreign)
write.table(ds, file="full_file_location_and_name")
```

Note: The `sep` option to `write.table()` can be used to change the default delimiter (space) to an arbitrary value.

1.2.5   Creating Excel spreadsheets

```
library(WriteXLS)
HELP = read.csv("http://www.amherst.edu/~nhorton/r2/datasets/help.csv")
WriteXLS("HELP", ExcelFileName="newhelp.xls")
```

Note: The `WriteXLS` package provides this functionality. It uses Perl (Practical extraction and report language, `http://www.perl.org`) and requires an external installation of Perl to function. After installing Perl, this requires running the operating system command `cpan -i Text::CSV_XS` at the command line.

1.2.6   Creating files for use by other packages

Example: 2.6.1

See also 1.2.8 (write XML).
```
library(foreign)
write.dta(ds, "filename.dta")
write.dbf(ds, "filename.dbf")
write.foreign(ds, "filename.dat", "filename.sas", package="SAS")
```

Note: Support for writing dataframes is provided in the `foreign` package. It is possible to write files directly in Stata format (see `write.dta()`) or DBF format (see `write.dbf()`) or create files with fixed fields as well as the code to read the file from within Stata, SAS, or SPSS using `write.foreign()`).

1.2.7   Creating HTML formatted output

```
library(prettyR)
htmlize("script.R", title="mytitle", echo=TRUE)
```

Note: The `htmlize()` function within the `prettyR` package can be used to produce HTML (hypertext markup language) from a script file (see A.2.1). The `cat()` function is used inside the script file (here denoted by `script.R`) to generate output. The `hwriter` package also supports writing R objects in HTML format. In addition, general HTML files can be created using the `markdown` package and the `markdownToHTML()` function; this can be integrated with the `knitr` package for reproducible analysis and is simplified in RStudio (11.3).

1.2.8 Creating XML datasets and output

The `XML` package provides support for writing XML files (see "Further resources").

1.3 Further resources

An introduction to data input and output can be found in [181]. Paul Murrell's *Introduction to Data Technologies* text [119] provides a comprehensive introduction to XML, SQL, and other related technologies and can be found at `http://www.stat.auckland.ac.nz/~paul/ItDT` (see also Nolan and Temple Lang [122]).

Chapter 2

Data management

This chapter reviews important data management tasks, including dataset structure, derived variables, and dataset manipulations. Along with functions available in base R, we demonstrate additional functions from the `dplyr`, `memisc`, `mosaic`, and `tidyr` packages.

2.1 Structure and metadata

2.1.1 Access variables from a dataset

The standard object to store data in R is the dataframe (see A.4.6), a rectangular collection of variables. Variables are generally stored as vectors. Variable references must contain the name of the object, which includes the variable, with certain exceptions.

```
with(ds, mean(x))
mean(ds$x)
```

Note: The `with()` and `within()` functions provide a way to access variables within a dataframe. In addition, the variables can be accessed directly using the $ operator. Many functions (e.g., `lm()`) allow specification of a dataset to be accessed using the `data` option.

The command `attach()` will make the variables within the named dataset available in the workspace, while `detach()` will remove them from the workspace (see also `conflicts()`). The Google R Style Guide [54] states that "the possibilities for creating errors when using `attach()` are numerous. Avoid it." We concur.

2.1.2 Names of variables and their types

Example: 2.6.1

```
str(ds)
```

Note: The command `sapply(ds, class)` will return the names and classes (e.g., numeric, integer, or character) of each variable within a dataframe, while running `summary(ds)` will provide an overview of the distribution of each column.

2.1.3   Values of variables in a dataset

Example: 2.6.2

```
print(ds)
```
or
```
View(ds)
```
or
```
edit(ds)
```
or
```
ds[1:10,]
ds[,2:3]
```

Note: The `print()` function lists the contents of the dataframe (or any other object), while the `View()` function opens a navigable window with a read-only view. The contents can be changed using the `edit()` function (this is not supported in the RStudio server version). Alternatively, any subset of the dataframe can be displayed on the screen using indexing, as in the final example. In `ds[1:10,]` the first 10 rows are displayed, while in `ds[,2:3]` the second and third variables. Variables can also be specified by name using a character vector index (see A.4.2). The `head()` function can be used to display the first (or, using `tail()`, last) values of a vector, dataset, or other object. Numbers will sometimes be displayed in scientific notation: the command `options(scipen=)` can be used to influence whether numeric values are displayed using fixed or exponential (scientific) notation.

2.1.4   Label variables

See also 2.2.19 (formatting variables).

Sometimes it is desirable to have a longer, more descriptive variable name. In general, we do not recommend using this feature, as it tends to complicate communication between data analysts and other readers of output.

```
comment(x) = "This is the label for the variable 'x'"
```

Note: The label for the variable can be extracted using `comment(x)` with no assignment or as `attributes(x)$comment`.

2.1.5   Add comment to a dataset or variable

Example: 2.6.1

To help facilitate proper documentation of datasets, it can be useful to provide some annotation or description.

```
comment(ds) = "This is a comment about the dataset"
```

Note: The `attributes()` function (see A.4.7) can be used to list all attributes, including any `comment()`, while the `comment()` function without an argument on the right-hand side will display the comment, if present.

2.2   Derived variables and data manipulation

This section describes the creation of new variables as a function of existing variables in a dataset.

2.2.1   Add derived variable to a dataset

Example: 6.6

```
library(dplyr)
ds = mutate(ds, newvar=myfunction(oldvar1, oldvar2, ...))
```
or
```
ds$newvar = with(ds, myfunction(oldvar1, oldvar2, ...))
```

Note: The routines in the `dplyr` package have been highly optimized, and often run dramatically faster than other options. In these equivalent examples, the new variable is added to the original dataframe. While care should be taken whenever dataframes are overwritten, this may be less risky because the addition of the variables is not connected with other changes.

2.2.2   Rename variables in a dataset

```
library(dplyr)
ds = rename(ds, new1=old1, new2=old2)
```
or
```
names(ds)[names(ds)=="old1"] = "new1"
names(ds)[names(ds)=="old2"] = "new2"
```
or
```
ds = within(ds, {new1 = old1; new2 = old2; rm(old1, old2)})
```

Note: The `rename()` function within the `dplyr` package provides a simple and efficient interface to rename variables in a dataframe. Alternatively, the `names()` function provides a list of names associated with an object (see A.4.6). The `edit()` function can be used to view names and edit values.

2.2.3   Create string variables from numeric variables

```
stringx = as.character(numericx)
typeof(stringx)
typeof(numericx)
```

Note: The `typeof()` function can be used to verify the type of an object; possible values include `logical`, `integer`, `double`, `complex`, `character`, `raw`, `list`, `NULL`, `closure` (function), `special`, and `builtin` (see A.4.7).

2.2.4   Create categorical variables from continuous variables

Examples: 2.6.3 and 7.10.6

```
newcat1 = (x >= cutpoint1) + ... + (x >= cutpointn)
```
or
```
newcat = cut(x, breaks=c(minval, cutpoint1, ..., cutpointn),
    labels=c("Cut1", "Cut2", ..., "Cutn"), right=FALSE)
```

Note: In the first implementation, each expression within parentheses is a logical test returning 1 if the expression is true, 0 if not true, and NA if x is missing. More information about missing value coding can be found in 11.4.4.1. The `cut()` function provides a more general framework (see also `cut_number()` from the `ggplot2` package).

2.2.5   Recode a categorical variable

A categorical variable may need to be recoded to have fewer levels (see also 6.1.3, changing reference category).

```
library(memisc)
newcat1=cases(
    "newval1"= oldcat==val1 | oldcat==val2,
    "newval2"= oldcat==valn)
```
or
```
tmpcat = oldcat
tmpcat[oldcat==val1] = newval1
tmpcat[oldcat==val2] = newval1
...
tmpcat[oldcat==valn] = newvaln
newcat = as.factor(tmpcat)
```

Note: The `cases()` function from the `memisc` package can be used to create the factor vector in one operation, by specifying the Boolean conditions. Alternatively, creating the variable can be undertaken in multiple steps. A copy of the old variable is first made, then multiple assignments are made for each of the new levels, for observations matching the condition inside the index (see A.4.2). In the final step, the categorical variable is coerced into a factor (class) variable.

2.2.6   Create a categorical variable using logic

Example: 2.6.3

Here we create a trichotomous variable `newvar`, which takes on a missing value if the continuous non-negative variable `oldvar` is less than 0, 0 if the continuous variable is 0, value 1 for subjects in group A with values greater than 0 but less than 50 and for subjects in group B with values greater than 0 but less than 60, or value 2 with values above those thresholds (more information about missing value coding can be found in 11.4.4.1).

```
library(memisc)
tmpvar = cases(
  "0" = oldvar==0,
  "1" = (oldvar>0 & oldvar<50 & group=="A") |
        (oldvar>0 & oldvar<60 & group=="B"),
  "2" = (oldvar>=50 & group=="A") |
        (oldvar>=60 & group=="B"))
```
or
```
tmpvar = rep(NA, length(oldvar))
tmpvar[oldvar==0] = 0
tmpvar[oldvar>0 & oldvar<50 & group=="A"] = 1
tmpvar[oldvar>0 & oldvar<60 & group=="B"] = 1
tmpvar[oldvar>=50 & group=="A"] = 2
tmpvar[oldvar>=60 & group=="B"] = 2
newvar = as.factor(tmpvar)
```

Note: Creating the variable is undertaken in multiple steps in the second approach. A vector of the correct length is first created containing missing values. Values are updated if they match the conditions inside the vector index (see A.4.2). Care needs to be taken in the comparison of `oldvar==0` if noninteger values are present (see 3.2.5).

The `cases()` function from the `memisc` package provides a straightforward syntax for derivations of this sort. The `%in%` operator can also be used to test whether a string is included in a larger set of possible values (see 2.2.11 and `help("%in%")`).

2.2.7 Create numeric variables from string variables

```
numericx = as.numeric(stringx)
typeof(stringx)
typeof(numericx)
```
or
```
stringf = factor(stringx)
numericx = as.numeric(stringf)
```

Note: The first set of code can be used when the string variable records numbers as character strings, and the code converts the storage type for these values. The second set of code can be used when the values in the string variable are arbitrary and may be awkward to enumerate for coding based on logical operations. The `typeof()` function can be used to verify the type of an object (see 2.2.3 and A.4.7).

2.2.8 Extract characters from string variables

```
get2through4 = substr(x, start=2, stop=4)
```

Note: The arguments to `substr()` specify the input vector, start character position, and end character position. The `stringr` package provides additional support for operations on character strings.

2.2.9 Length of string variables

```
len = nchar(stringx)
```

Note: The `nchar()` function returns a vector of lengths of each of the elements of the string vector given as argument, as opposed to the `length()` function (2.3.4) that returns the number of elements in a vector. The `stringr` package provides additional support for operations on character strings.

2.2.10 Concatenate string variables

```
newcharvar = paste(x1, " VAR2 ", x2, sep="")
```

Note: The above R code creates a character variable `newcharvar` containing the character vector X_1 (which may be coerced from a numeric object) followed by the string " VAR2 " then the character vector X_2. The `sep=""` option leaves no additional separation character between these three strings.

2.2.11   Set operations

```
newengland = c("MA", "CT", "RI", "VT", "ME", "NH")
"NY" %in% newengland
"MA" %in% newengland
```

Note: The first statement would return FALSE, while the second one would return TRUE. The %in% operator also works with numeric vectors (see help("%in%")). Vector functions for set-like operations include union(), setdiff(), setequal(), intersect(), unique(), duplicated(), and match().

2.2.12   Find strings within string variables

Example: 7.10.9

```
matches = grep("pat", stringx)
positions = regexpr("pat", stringx)
```

```
> x = c("abc", "def", "abcdef", "defabc")
> grep("abc", x)
[1] 1 3 4
> regexpr("abc", x)
[1]  1 -1  1  4
attr(,"match.length")
[1]  3 -1  3  3
attr(,"useBytes")
[1] TRUE
> regexpr("abc", x) < 0
[1] FALSE  TRUE FALSE FALSE
```

Note: The function grep() returns a list of elements in the vector given by stringx that match the given pattern, while the regexpr() function returns a numeric list of starting points in each string in the list (with −1 if there was no match). Testing positions < 0 generates a vector of binary indicator of matches (TRUE=no match, FALSE=a match).

The regular expressions available within grep and other related routines are quite powerful. As an example, Boolean OR expressions can be specified using the | operator. A comprehensive description of these operators can be found using help(regex). Additional support for operations on character vectors can be found in the **stringr** package.

2.2.13   Find approximate strings

```
agrep(pattern, string, max.distance=n)
```

Note: The support within the agrep() function is more rudimentary: it calculates the Levenshtein edit distance (total number of insertions, deletions, and substitutions required to transform one string into another) and it returns the indices of the elements of the second argument that are within n edits of pattern (see 2.2.12). By default, the threshold is 10% of the pattern length.

```
> x = c("I ask a favour", "Pardon my error", "You are my favorite")
> agrep("favor", x, max.distance=1)
[1] 1 3
```

2.2.14 Replace strings within string variables

Example: 12.2

```
newstring = gsub("oldpat", "newpat", oldstring)
```
or
```
x = "oldpat123"
substr(x, start=1, stop=6) = "newpat"
```

2.2.15 Split strings into multiple strings

```
strsplit(string, splitchar)
```

Note: The function `strsplit()` returns a list, each element of which is a vector containing the parts of the input, split at each occurrence of `splitchar`. If the input is a single character string, this is a list of one vector. If `split` is the null string, then the function returns a list of vectors of single characters.

```
> x = "this is a test"
> strsplit(x, " ")
[[1]]
[1] "this" "is"   "a"    "test"
> strsplit(x,"")
[[1]]
 [1] "t" "h" "i" "s" " " "i" "s" " " "a" " " "t" "e" "s" "t"
```

2.2.16 Remove spaces around string variables

```
noleadortrail = sub(' +$', '', sub('^ +', '', stringx))
```

Note: The arguments to `sub()` consist of a regular expression, a substitution value, and a vector. In the first step, leading spaces are removed, then a separate call to `sub()` is used to remove trailing spaces (in both cases replacing the spaces with the null string). If instead of spaces all trailing whitespaces (e.g., tabs, space characters) should be removed, the regular expression ' +$' should be replaced by '[[:space:]]+$'.

2.2.17 Convert strings from upper to lower case

```
lowercasex = tolower(x)
```
or
```
lowercasex = chartr("ABCDEFGHIJKLMNOPQRSTUVWXYZ",
      "abcdefghijklmnopqrstuvwxzy", x)
```

Note: The `toupper()` function can be used to convert to upper case. Arbitrary translations from sets of characters can be made using the `chartr()` function. The `iconv()` supports more complex encodings (e.g., from ASCII to other languages).

2.2.18 Create lagged variable

A lagged variable has the value of that variable in a previous row (typically the immediately previous one) within that dataset. The value of lag for the first observation will be missing (see 11.4.4.1).

```
lag1 = c(NA, x[1:(length(x)-1)])
```

Note: This expression creates a one-observation lag, with a missing value in the first position, and the first through second-to-last observation for the remaining entries (see `lag()`). Here we demonstrate how to write a function to create lags of more than one observation.

```
lagk = function(x, k) {
    len = length(x)
    if (!floor(k)==k) {
        cat("k must be an integer")
    } else if (k<1 | k>(len-1)) {
        cat("k must be between 1 and length(x)-1")
    } else {
        return(c(rep(NA, k), x[1:(len-k)]))
    }
}

> lagk(1:10, 5)
 [1] NA NA NA NA NA  1  2  3  4  5
```

2.2.19   Formatting values of variables

Example: 6.6.2

See also 2.1.4 (labelling variables).

Sometimes it is useful to display category names that are more descriptive than variable names. In general, we do not recommend using this feature (except potentially for graphical output), as it tends to complicate communication between data analysts and other readers of output. In this example, character labels are associated with a numeric variable (0=Control, 1=Low Dose, and 2=High Dose).

```
> x = c(0, 0, 1, 1, 2); x
[1] 0 0 1 1 2
> x = factor(x, 0:2, labels=c("Control", "Low Dose", "High Dose")); x
[1] Control   Control   Low Dose  Low Dose  High Dose
Levels: Control Low Dose High Dose
```

Note: The `rownames()` function can be used to associate a variable with an identifier (which is by default the observation number). As an example, this can be used to display the name of a region with the value taken by a particular variable measured in that region. The `setNames()` function can also be used to set the names on an object.

2.2.20   Perl interface

Perl is a high-level general-purpose programming language [154]. The `RSPerl` package provides a bidirectional interface between Perl and R.

2.2.21   Accessing databases using SQL

Example: 12.7

Structured Query Language (SQL) is a flexible language for accessing and modifying databases, data warehouses, and distributed systems. These interfaces are particularly useful when analyzing large datasets, since databases are highly optimized for certain complex operations such as merges (joins).

The RODBC, RMySQL, and RSQLite packages provide access to SQL within R [135]. The dplyr package provides a grammar of data manipulation that is optimized for dataframes, datatables, and databases. The RMongo package provides an interface to NoSQL Mongo databases (http://www.mongodb.org). Access and analysis of a large external database is demonstrated in 12.7.

Selections and other operations can be made on dataframes using an SQL-interface with the sqldf package.

2.3 Merging, combining, and subsetting datasets

A common task in data analysis involves the combination, collation, and subsetting of datasets. In this section, we review these techniques for a variety of situations.

2.3.1 Subsetting observations

Example: 2.6.4

```
library(dplyr)
smallds = filter(ds, x==1)
```
or
```
smallds = ds[x==1,]
```
or
```
smallds = subset(ds, x==1)
```

Note: Each example creates a subset of a dataframe consisting of observations where $X = 1$. In addition, many functions allow specification of a subset=expression option to carry out a procedure on observations that match the expression (see also slice() in the dplyr package). The routines in the dplyr package have been highly optimized, and often run dramatically faster than other options.

2.3.2 Drop or keep variables in a dataset

Example: 2.6.1

It is often desirable to prune extraneous variables from a dataset to simplify analyses. This can be done by specifying a set to keep or a set to drop.

```
library(dplyr)
narrow = select(ds, x1, xk)
```
or
```
narrow = ds[,c("x1", "xk")]
```
or
```
narrow = subset(ds, select = c(x1, xk))
```

Note: The examples create a new dataframe consisting of the variables x1 and xk. Each approach allows the specification of a set of variables to be excluded. The routines in the dplyr package have been highly optimized, and often run dramatically faster than other options.

More sophisticated ways of listing the variables to be kept are available. The dplyr package includes functions starts_with(), ends_with(), contains(), matches(), num_range(), and one_of. In base R, the command ds[,grep("x1|^pat", names(ds))] would keep x1 and all variables starting with pat (see 2.2.12).

2.3.3   Random sample of a dataset

It is sometimes useful to sample a subset (here quantified as **nsamp**) of observations without replacement from a larger dataset (see random number seed, 3.1.3).

```
library(mosaic)
newds = resample(ds, size=nsamp, replace=FALSE)
```
or
```
newds = ds[sample(nrow(ds), size=nsamp),]
```

Note: By default, the `resample()` function in the `mosaic` package creates a sample without replacement from a dataframe or vector (the built-in `sample()` function cannot directly sample a dataframe). The `replace=TRUE` option can be used to override this (e.g., when bootstrapping, see 11.4.3). In the second example, the `sample()` function from base R is used to get a random selection of row numbers, in conjunction with the `nrow()` function, which returns the number of rows.

2.3.4   Observation number

```
> library(dplyr)
> ds = data.frame(y = c("abc", "def", "ghi"))
> ds = mutate(ds, id = 1:nrow(ds))
> ds
    y id
1 abc  1
2 def  2
3 ghi  3
```

Note: The `nrow()` function returns the number of rows in a dataframe. Here, it is used in conjunction with the `:` operator (4.1.3) to create a vector with the integers from 1 to the sample size. These can then be added to the dataframe, as shown, or might be used as row labels (see `names()`). The `length()` function returns the number of elements in a vector, while the `dim()` function returns the dimension (number of rows and columns) for a dataframe (A.4.6).

2.3.5   Keep unique values

See also 2.3.6 (duplicated values).

```
uniquevalues = unique(x)
uniquevalues = unique(data.frame(x1, ..., xk))
```

Note: The `unique()` function returns each of the unique values represented by the vector or dataframe denoted by `x` (see also `distinct()` in the `dplyr` package).

2.3.6   Identify duplicated values

See also 2.3.5 (unique values).

```
duplicated(x)
```

Note: The `duplicated()` function returns a logical vector indicating a replicated value. Note that the first occurrence is not a replicated value. Thus `duplicated(c(1,1))` returns **FALSE TRUE**.

2.3.7 Convert from wide to long (tall) format

Example: 7.10.9

Data are often found in a different shape than that required for analysis. One example of this is commonly found in longitudinal measures studies. In this setting it is convenient to store the data in a wide or multivariate format with one line per observation, containing typically subject-invariant factors (e.g., gender), as well as a column for each repeated outcome. An example is given below.

```
id female inc80 inc81 inc82
1    0    5000  5500  6000
2    1    2000  2200  3300
3    0    3000  2000  1000
```

Here, the income for 1980, 1981, and 1982 are included in one row for each id.

In contrast, tools for repeated measures analyses (7.4.2) typically require a row for each repeated outcome, as demonstrated below.

```
id year female inc
1  80   0      5000
1  81   0      5500
1  82   0      6000
2  80   1      2000
2  81   1      2200
2  82   1      3300
3  80   0      3000
3  81   0      2000
3  82   0      1000
```

In this section and in 2.3.8, we show how to convert between these two forms of this example data.

```
library(dplyr); library(tidyr)
long = ds %>%
  gather(year, inc, inc80:inc82) %>%
  mutate(year = extract_numeric(year)) %>%
  arrange(id, year)
```

Note: The `gather()` function in the `tidyr` package takes a dataframe, a "key" (in this case `year`), "value" (in this case `inc`), and list of variables as arguments, and transposes the dataset. The "key" will be the name of a new variable containing the names of the variables in the list. The "value" will be the name of a new variable containing the values in the variables in the list. For each row of the original dataset, the output dataset will contain a row for each of the variables in the list, so that each variable–value pair appears exactly once in both datasets, but in the output dataset, all the values are in the "value" column. The non-listed variables will be repeated in each row. Here, the output from this operation is piped (see A.5.3) to the `mutate()` function, which extracts the numeric value from the year variable. Finally, the `arrange()` function reorders the resulting dataframe by id and year.

2.3.8 Convert from long (tall) to wide format

See also 2.3.7 (reshape from wide to tall).

```
library(dplyr); library(tidyr)
wide = long %>%
  mutate(year=paste("inc", year, sep="")) %>%
  spread(year, inc)
```

Note: This example assumes that the dataset `long` has repeated measures on `inc` for subject `id` determined by the variable `year`. The call to mutate is needed to prepend the string `"inc"` to the newly created variables, then pipe (see A.5.3) the resulting output to the `spread()` function (which is the inverse of the `gather()` function: see 2.3.7).

2.3.9 Concatenate and stack datasets

```
newds = rbind(ds1, ds2)
```

Note: The result of `rbind()` is a dataframe with as many rows as the sum of rows in `ds1` and `ds2`. Dataframes given as arguments to `rbind()` must have the same column names. The similar `cbind()` function makes a dataframe with as many columns as the sum of the columns in the input objects. A similar function (`c()`) operates on vectors.

2.3.10 Sort datasets

Example: 2.6.4

```
library(dplyr)
sortds = arrange(ds, x1, x2, ..., xk)
```
or
```
sortds = ds[with(ds, order(x1, x2, ..., xk)),]
```
Note: The `arrange()` function within the `dplyr` package provides a way to sort the rows within dataframes. The `desc()` function can be applied to one of the arguments to sort in a descending fashion. The R command `sort()` can also be used to sort a vector, while `order()` can be used to sort dataframes by selecting a new permutation of order for the rows. The `decreasing` option can be used to change the default sort order (for all variables). As an alternative, a numeric variable can be reversed by specifying `-x1` instead of `x1`. The routines in the `dplyr` package have been highly optimized, and typically run dramatically faster than other options.

2.3.11 Merge datasets

Example: 7.10.11

Merging datasets is commonly required when data on single units are stored in multiple tables or datasets. We consider a simple example where variables `id, year, female`, and `inc` are available in one dataset, and variables `id` and `maxval` in a second. For this simple example, with the first dataset given as:

```
id year female inc
1  80   0      5000
1  81   0      5500
1  82   0      6000
2  80   1      2000
2  81   1      2200
```

```
2  82  1      3300
3  80  0      3000
3  81  0      2000
3  82  0      1000
```

and the second given below.

```
id maxval
2  2400
1  1800
4  1900
```

The desired merged dataset would look like the following (an outer join, where all observations are included if they are present in either of the dataframes).

```
   id year female  inc maxval
1   1   81      0 5500   1800
2   1   80      0 5000   1800
3   1   82      0 6000   1800
4   2   82      1 3300   2400
5   2   80      1 2000   2400
6   2   81      1 2200   2400
7   3   82      0 1000     NA
8   3   80      0 3000     NA
9   3   81      0 2000     NA
10  4   NA     NA   NA   1900
```

```
merged2 = merge(ds2, ds1, by="id", all=TRUE)
```
or
```
library(dplyr)
merged = union(left_join(ds1, ds2, by="id"),
      left_join(ds2, ds1, by="id"))
```

Note: The `merge()` function allows outer joins with the `all=TRUE` option, and left or right joins with `all.x` and `all.y`, respectively. While `dplyr` is generally more flexible and faster, it does not directly support an outer join function. This can be emulated by use of the `union()` function and two calls to `left_join()`. Multiple variables can be specified in the by option. Other types of merges (e.g., inner joins) are supported. The command `inner_join(ds1, ds2, by="id")` would yield the same dataset with no missing values.

2.4 Date and time variables

The standard date functions in R return a `Date` class that represents the number of days since January 1, 1970. The `chron` and `lubridate` packages also provide support for manipulations of dates.

2.4.1 Create date variable

See also 1.1.9 (read more complex files).

```
dayvar = as.Date("2016-04-29")
todays_date = as.Date(Sys.time())
```

Note: The return value of `as.Date()` is a `Date` class object. If converted to numeric `dayvar`, it represents the number of days between January 1, 1970 and April 29, 2016, while `todays_date` is the integer number of days since January 1, 1970 (see `ISOdate()`).

2.4.2 Extract weekday

```
wkday = weekdays(datevar)
```

Note: `wkday` contains a string with the name of the weekday of the `Date` object.

2.4.3 Extract month

```
monthval = months(datevar)
```

Note: The function `months()` returns a string with the name of the month of the `Date` object.

2.4.4 Extract year

```
yearval = substr(as.POSIXct(datevar), 1, 4)
```

Note: The `as.POSIXct()` function returns a string representing the date, with the first four characters corresponding to the year.

2.4.5 Extract quarter

```
qtrval = quarters(datevar)
```

Note: The function `quarters()` returns a string representing the quarter of the year (e.g., `"Q1"` or `"Q2"`) given by the `Date` object.

2.4.6 Create time variable

Example: 12.4.2

See also 4.3.1 (timing commands)

```
> arbtime = as.POSIXlt("2016-04-29 17:15:45 NZDT")
> arbtime
[1] "2016-04-29 17:15:45"
> now = Sys.time()
> now
[1] "2016-04-01 10:12:11 EST"
```

Note: The objects `arbtime` and `now` can be compared with the subtraction operator to monitor elapsed time.

2.5 Further resources

Comprehensive introductions to data management in R can be found in [181]. Hadley Wickham's `dplyr` package [194] provides a number of useful data management routines that can efficiently operated on dataframes, datatables, and databases. The `tidyr` package [192] facilitates data cleaning and preparation.

2.6 Examples

To help illustrate the tools presented in this and related chapters, we apply many of the entries to the HELP RCT data. The code can be downloaded from `http://www.amherst.edu/~nhorton/r2/examples`.

2.6.1 Data input and output

We begin by reading the dataset (1.1.4), keeping only the variables that are needed (2.3.2).

```
> options(digits=3)
> options(width=72) # narrow output
> ds = read.csv("http://www.amherst.edu/~nhorton/r2/datasets/help.csv")
> library(dplyr)
> newds = select(ds, cesd, female, i1, i2, id, treat, f1a, f1b, f1c, f1d,
    f1e, f1f, f1g, f1h, f1i, f1j, f1k, f1l, f1m, f1n, f1o, f1p, f1q, f1r,
    f1s, f1t)
```

We can then show a summary of the dataset.

```
> names(newds)
 [1] "cesd"   "female" "i1"     "i2"     "id"     "treat"  "f1a"
 [8] "f1b"    "f1c"    "f1d"    "f1e"    "f1f"    "f1g"    "f1h"
[15] "f1i"    "f1j"    "f1k"    "f1l"    "f1m"    "f1n"    "f1o"
[22] "f1p"    "f1q"    "f1r"    "f1s"    "f1t"
> str(newds[,1:10]) # structure of the first 10 variables
'data.frame': 453 obs. of  10 variables:
 $ cesd  : int  49 30 39 15 39 6 52 32 50 46 ...
 $ female: int  0 0 0 1 0 1 1 0 1 0 ...
 $ i1    : int  13 56 0 5 10 4 13 12 71 20 ...
 $ i2    : int  26 62 0 5 13 4 20 24 129 27 ...
 $ id    : int  1 2 3 4 5 6 7 8 9 10 ...
 $ treat : int  1 1 0 0 0 1 0 1 0 1 ...
 $ f1a   : int  3 3 3 0 3 1 3 1 3 2 ...
 $ f1b   : int  2 2 2 0 0 0 1 1 2 3 ...
 $ f1c   : int  3 0 3 1 3 1 3 2 3 3 ...
 $ f1d   : int  0 3 0 3 3 3 1 3 1 0 ...
```

```
> summary(newds[,1:10]) # summary of the first 10 variables
      cesd            female            i1              i2
 Min.   : 1.0    Min.   :0.000   Min.   :  0.0   Min.   :  0.0
 1st Qu.:25.0    1st Qu.:0.000   1st Qu.:  3.0   1st Qu.:  3.0
 Median :34.0    Median :0.000   Median : 13.0   Median : 15.0
```

```
 Mean    :32.8   Mean    :0.236   Mean    : 17.9   Mean    : 22.6
 3rd Qu.:41.0    3rd Qu.:0.000    3rd Qu.: 26.0    3rd Qu.: 32.0
 Max.    :60.0   Max.    :1.000   Max.    :142.0   Max.    :184.0
        id              treat              f1a              f1b
 Min.    :  1    Min.    :0.000   Min.    :0.00    Min.    :0.00
 1st Qu.:119     1st Qu.:0.000    1st Qu.:1.00     1st Qu.:0.00
 Median :233     Median :0.000    Median :2.00     Median :1.00
 Mean    :233    Mean    :0.497   Mean    :1.63    Mean    :1.39
 3rd Qu.:348     3rd Qu.:1.000    3rd Qu.:3.00     3rd Qu.:2.00
 Max.    :470    Max.    :1.000   Max.    :3.00    Max.    :3.00
        f1c              f1d
 Min.    :0.00   Min.    :0.00
 1st Qu.:1.00    1st Qu.:0.00
 Median :2.00    Median :1.00
 Mean    :1.92   Mean    :1.56
 3rd Qu.:3.00    3rd Qu.:3.00
 Max.    :3.00   Max.    :3.00
```

Displaying the first few rows of data can give a more concrete sense of what is in the dataset.

```
> head(newds, n=3)
  cesd female i1 i2 id treat f1a f1b f1c f1d f1e f1f f1g f1h f1i f1j
1   49      0 13 26  1     1   3   2   3   0   2   3   3   0   2   3
2   30      0 56 62  2     1   3   2   0   3   3   2   0   0   3   0
3   39      0  0  0  3     0   3   2   3   0   2   2   1   3   2   3
  f1k f1l f1m f1n f1o f1p f1q f1r f1s f1t
1   3   0   1   2   2   2   2   3   3   2
2   3   0   0   3   0   0   0   2   0   0
3   1   0   1   3   2   0   0   3   2   0
```

Saving the dataset in native format (1.2.3) will ease future access. We also add a comment (2.1.5) to help later users understand what is in the dataset.

```
> comment(newds) = "HELP baseline dataset"
> comment(newds)
[1] "HELP baseline dataset"
> save(ds, file="savedfile")
```

Saving it in a foreign format (1.1.8), say Microsoft Excel or comma-separated value format, will allow access to other tools for analysis and display.

```
> write.csv(ds, file="ds.csv")
```

Creating data in SAS format from R can be useful; note that the R code below generates an ASCII dataset and a SAS command file to read it into SAS.

```
> library(foreign)
> write.foreign(newds, "file.dat", "file.sas", package="SAS")
```

2.6.2 Data display

We begin by consideration of the CESD (Center for Epidemiologic Statistics) measure of depressive symptoms for this sample at baseline. The indexing mechanisms (see A.4.2) are helpful in extracting subsets of a vector.

```
> with(newds, cesd[1:10])
 [1] 49 30 39 15 39  6 52 32 50 46
> with(newds, head(cesd, 10))
 [1] 49 30 39 15 39  6 52 32 50 46
```

It may be useful to know what high values there are.

```
> with(newds, cesd[cesd > 56])
[1] 57 58 57 60 58 58 57
```

```
> library(dplyr)
> filter(newds, cesd > 56) %>% select(id, cesd)
   id cesd
1  71   57
2 127   58
3 200   57
4 228   60
5 273   58
6 351   58
7  13   57
```

In the first example, we subset to display the values matching the condition. In the second example, the `filter()` function from the `dplyr` package is used to subset the rows, then `select()` is used to display a subset of columns (these are connected using the `%>%` operator, see A.5.3).

In a similar fashion, it may be useful to examine the observations with the lowest values.

```
> with(newds, sort(cesd)[1:4])
[1] 1 3 3 4
> with(newds, which.min(cesd))
[1] 199
```

2.6.3 Derived variables and data manipulation

Suppose the dataset arrived with only the individual CESD questions and not the sum. We would need to create the CESD score. Note that there are four questions which are asked "backwards," meaning that high values of the response are counted for fewer points.[1] We'll approach the recoding of the flipped questions by reading the CESD items into a new object. To demonstrate other tools, we'll also see if there's any missing data (11.4.4.1) and generate some other statistics about the question responses.

---

[1] This follows from the coding instructions found at `http://www.amherst.edu/~nhorton/r2/cesd.pdf`.

```
> library(mosaic)
> tally(~ is.na(f1g), data=newds)

 TRUE FALSE
    1   452
> favstats(~ f1g, data=newds)
 min Q1 median Q3 max mean   sd   n missing
   0  1      2  3   3 1.73  1.1 452       1
```

We utilize the `tally()` and `favstat()` functions from the `mosaic` package to display the distribution and number of missing values.

Now we're ready to create the score. We'll generate the sum of the non-missing items, which effectively imputes 0 for the missing values, as well as a version that imputes the mean of the observed values instead.

```
> # reverse code f1d, f1h, f1l and f1p
> cesditems = with(newds, cbind(f1a, f1b, f1c, (3 - f1d), f1e, f1f, f1g,
      (3 - f1h), f1i, f1j, f1k, (3 - f1l), f1m, f1n, f1o, (3 - f1p),
      f1q, f1r, f1s, f1t))
> nmisscesd = apply(is.na(cesditems), 1, sum)
> ncesditems = cesditems
> ncesditems[is.na(cesditems)] = 0
> newcesd = apply(ncesditems, 1, sum)
> imputemeancesd = 20/(20-nmisscesd)*newcesd
```

It is prudent to review the results when deriving variables. We'll check our re-created CESD score against the one that came with the dataset. To ensure that missing data has been correctly coded, we print the subjects with any missing questions.

```
> data.frame(newcesd, newds$cesd, nmisscesd, imputemeancesd)[nmisscesd>0,]
    newcesd newds.cesd nmisscesd imputemeancesd
4        15         15         1           15.8
17       19         19         1           20.0
87       44         44         1           46.3
101      17         17         1           17.9
154      29         29         1           30.5
177      44         44         1           46.3
229      39         39         1           41.1
```

The output shows that the original dataset was created with unanswered questions counted as if they had been answered with a zero. This conforms to the instructions provided with the CESD, but might be questioned on theoretical grounds.

It is often necessary to create a new variable using logic (2.2.6). In the HELP study, many subjects reported extreme amounts of drinking (as the baseline measure was taken while they were in detox). Here, an ordinal measure of alcohol consumption (abstinent, moderate, high-risk) is created using information about average consumption per day in the 30 days prior to detox (i1, measured in standard drink units) and maximum number of drinks per day in the 30 days prior to detox (i2). The number of drinks required for each category differs for men and women according to NIAAA guidelines for physicians [121].

```
> library(dplyr)
> library(memisc)
> newds = mutate(newds, drinkstat=
    cases(
      "abstinent" = i1==0,
      "moderate" = (i1>0 & i1<=1 & i2<=3 & female==1) |
                   (i1>0 & i1<=2 & i2<=4 & female==0),
      "highrisk" = ((i1>1 | i2>3) & female==1) |
                   ((i1>2 | i2>4) & female==0)))
```

Again we will double check our variable creation. Here we display the observations in the last 6 rows of the data.

```
> library(dplyr)
> tmpds = select(newds, i1, i2, female, drinkstat)
> tmpds[365:370,]
    i1 i2 female drinkstat
365  6 24      0  highrisk
366  6  6      0  highrisk
367  0  0      0 abstinent
368  0  0      1 abstinent
369  8  8      0  highrisk
370 32 32      0  highrisk
```

It is also useful to focus such checks on a subset of observations. Here we show the drinking data for moderate female drinkers.

```
> library(dplyr)
> filter(tmpds, drinkstat=="moderate" & female==1)
  i1 i2 female drinkstat
1  1  1      1  moderate
2  1  3      1  moderate
3  1  2      1  moderate
4  1  1      1  moderate
5  1  1      1  moderate
6  1  1      1  moderate
7  1  1      1  moderate
```

Basic data description is an early step in analysis. Here we calculate relevant summaries of drinking and gender.

```
> library(gmodels)
> with(tmpds, CrossTable(drinkstat))


   Cell Contents
|-----------------------|
|                     N |
|         N / Table Total |
|-----------------------|
```

```
Total Observations in Table:  453


              | abstinent |  moderate | highrisk |
              |-----------|-----------|----------|
              |        68 |        28 |      357 |
              |     0.150 |     0.062 |    0.788 |
              |-----------|-----------|----------|
```

```
> with(tmpds, CrossTable(drinkstat, female,
    prop.t=FALSE, prop.c=FALSE, prop.chisq=FALSE))


   Cell Contents
|-----------------------|
|                     N |
|         N / Row Total |
|-----------------------|


Total Observations in Table:  453


              | female
   drinkstat |         0 |         1 | Row Total |
-------------|-----------|-----------|-----------|
   abstinent |        42 |        26 |        68 |
             |     0.618 |     0.382 |     0.150 |
-------------|-----------|-----------|-----------|
    moderate |        21 |         7 |        28 |
             |     0.750 |     0.250 |     0.062 |
-------------|-----------|-----------|-----------|
    highrisk |       283 |        74 |       357 |
             |     0.793 |     0.207 |     0.788 |
-------------|-----------|-----------|-----------|
Column Total |       346 |       107 |       453 |
-------------|-----------|-----------|-----------|
```

To display gender more clearly, we create a new character variable. Note that as for other objects in R, quoted strings are case sensitive.

```
> newds = transform(newds,
    gender=factor(female, c(0,1), c("Male","Female")))
```

```
> tally(~ female + gender, margin=FALSE, data=newds)
      gender
female Male Female
     0  346      0
     1    0    107
```

2.6.4   Sorting and subsetting datasets

It is often useful to sort datasets (2.3.10) by the order of a particular variable (or variables).
Here we sort by CESD and drinking.

```
> library(dplyr)
> newds = arrange(ds, cesd, i1)
> newds[1:5, c("cesd", "i1", "id")]
  cesd i1  id
1    1  3 233
2    3  1 139
3    3 13 418
4    4  4 251
5    4  9  95
```

It is sometimes necessary to create data that is a subset (2.3.1) of other data. Here we make
a dataset that only includes female subjects. First, we create the subset and calculate a
summary value in the resulting dataset.

```
> library(dplyr)
> females = filter(ds, female==1)
> with(females, mean(cesd))
[1] 36.9
> # an alternative approach
> mean(ds$cesd[ds$female==1])
[1] 36.9
```

To test the subsetting, we then display the mean for both genders.

```
> with(ds, tapply(cesd, female, mean))
   0    1
31.6 36.9
> library(mosaic)
> mean(cesd ~ female, data=ds)
   0    1
31.6 36.9
```

Chapter 3

Statistical and mathematical functions

This chapter reviews key statistical, probability, mathematical, and matrix functions.

3.1 Probability distributions and random number generation

Quantiles and cumulative distribution values can be calculated easily within R. Random variables are commonly needed for simulation and analysis. These can be generated for a large number of distributions.

A seed can be specified for the random number generator. This is important to allow replication of results (e.g., while testing and debugging). Information about random number seeds can be found in 3.1.3.

Table 3.1 summarizes support for quantiles, cumulative distribution functions, and random numbers. More information on probability distributions can be found in the CRAN probability distributions task view (`http://cran.r-project.org/web/views/Distributions.html`).

3.1.1 Probability density function

Example: 3.4.1

Here we use the normal distribution as an example; others are shown in Table 3.1 (**p. 34**).

```
y = pnorm(1.96, mean=0, sd=1)
```

Note: This calculates the probability that the random variable is less than the first argument. The `xpnorm()` function within the **mosaic** package provides a graphical display.

3.1.2 Quantiles of a probability density function

Example: 4.2

Similar syntax is used for a variety of distributions. Here we use the normal distribution as an example; others are shown in Table 3.1 (**p. 34**).

```
y = qnorm(.975, mean=0, sd=1)
```

Table 3.1: Quantiles, probabilities, and pseudo-random number generation: available distributions.

Distribution	R DISTNAME
Beta	beta
Beta-binomial	betabin*
Beta-normal	betanorm*
binomial	binom
Cauchy	cauchy
chi-square	chisq
exponential	exp
F	f
gamma	gamma
geometric	geom
hypergeometric	hyper
inverse normal	inv.gaussian*
Laplace	alap*
logistic	logis
lognormal	lnorm
negative binomial	nbinom
normal	norm
Poisson	pois
Student's t	t
Uniform	unif
Weibull	weibull

Note: Prepend d to the command to compute density functions of a distribution dDISTNAME(xvalue, parm1, ..., parmn), p for the cumulative distribution function, pDISTNAME(xvalue, parm1, ..., parmn), q for the quantile function qDISTNAME(prob, parm1, ..., parmn), and r to generate random variables rDISTNAME(nrand, parm1, ..., parmn), where in the last case a vector of nrand values is the result.
* The betabinom(), betanorm(), inv.gaussian(), and alap() (Laplace) families of distributions are available using the VGAM package.

3.1.3 Setting the random number seed

Example: 12.1.3

The default random number seed is based on the system clock. To generate a replicable series of variates, first run set.seed(seedval) where seedval is a single integer for the default Mersenne–Twister random number generator.

```
set.seed(42)
set.seed(Sys.time())
```

Note: More information can be found using help(.Random.seed).

3.1.4 Uniform random variables

Example: 10.1.1

```
x = runif(n, min=0, max=1)
```

Note: The arguments specify the number of variables to be created and the range over which they are distributed.

3.1.5 Multinomial random variables

```
library(Hmisc)
x = rMultinom(matrix(c(p1, p2, ..., pr), 1, r), n)
```

Note: The function `rMultinom()` from the `Hmisc` package allows the specification of the desired multinomial probabilities ($\sum_r p_r = 1$) as a $1 \times r$ matrix. The final parameter is the number of variates to be generated (see also `rmultinom()` in the `stats` package).

3.1.6 Normal random variables

Example: 3.4.1

```
x1 = rnorm(n)
x2 = rnorm(n, mean=mu, sd=sigma)
```

Note: The arguments specify the number of variables to be created and (optionally) the mean and standard deviation (default $\mu = 0$ and $\sigma = 1$).

3.1.7 Multivariate normal random variables

For the following, we first create a 3×3 covariance matrix. Then we generate 1000 realizations of a multivariate normal vector with the appropriate correlation or covariance.

```
library(MASS)
mu = rep(0, 3)
Sigma = matrix(c(3, 1, 2,
                 1, 4, 0,
                 2, 0, 5), nrow=3)
xvals = mvrnorm(1000, mu, Sigma)
apply(xvals, 2, mean)
```
or
```
rmultnorm = function(n, mu, vmat, tol=1e-07)
# a function to generate random multivariate Gaussians
{
   p = ncol(vmat)
   if (length(mu)!=p)
      stop("mu vector is the wrong length")
   if (max(abs(vmat - t(vmat))) > tol)
      stop("vmat not symmetric")
   vs = svd(vmat)
   vsqrt = t(vs$v %*% (t(vs$u) * sqrt(vs$d)))
   ans = matrix(rnorm(n * p), nrow=n) %*% vsqrt
   ans = sweep(ans, 2, mu, "+")
   dimnames(ans) = list(NULL, dimnames(vmat)[[2]])
   return(ans)
}
xvals = rmultnorm(1000, mu, Sigma)
apply(xvals, 2, mean)
```

Note: The returned object `xvals`, of dimension 1000×3, is generated from the variance–covariance matrix denoted by `Sigma`, which has first row and column (3,1,2). An arbitrary mean vector can be specified using the `c()` function.

Several techniques are illustrated in the definition of the `rmultnorm` function. The first lines test for the appropriate arguments and return an error if the conditions are not satisfied. The singular value decomposition (see 3.3.15) is carried out on the variance–covariance matrix, and the `sweep` function is used to transform the univariate normal random variables generated by `rnorm` to the desired mean and covariance. The `dimnames()` function applies the existing names (if any) for the variables in `vmat`, and the result is returned.

3.1.8   Truncated multivariate normal random variables

See also 4.1.1.

```
library(tmvtnorm)
x = rtmvnorm(n, mean, Sigma, lower, upper)
```

Note: The arguments specify the number of variables to be created, the mean, the covariance matrix, and vectors of the lower and upper truncation values.

3.1.9   Exponential random variables

```
x = rexp(n, rate=lambda)
```

Note: The arguments specify the number of variables to be created and (optionally) the inverse of the mean (default $\lambda = 1$).

3.1.10   Other random variables

Example: 3.4.1

The list of probability distributions supported within R can be found in Table 3.1, **page 34**. In addition to these distributions, the inverse probability integral transform can be used to generate arbitrary random variables with invertible cumulative density function F (exploiting the fact that $F^{-1} \sim U(0,1)$). As an example, consider the generation of random variates from an exponential distribution with rate parameter λ, where $F(X) = 1 - \exp(-\lambda X) = U$. Solving for X yields $X = -\log(1-U)/\lambda$. If we generate a Uniform(0,1) variable, we can use this relationship to generate an exponential with the desired rate parameter.

```
lambda = 2
expvar = -log(1-runif(1))/lambda
```

3.2   Mathematical functions

3.2.1   Basic functions

See also 2.2 (derived variables) and 2.2.11 (sets).

```
minx = min(x)
maxx = max(x)
meanx = mean(x)
modx = x1 %% x2
stddevx = sd(x)
absolutevaluex = abs(x)
squarerootx = sqrt(x)
etothex = exp(x)
xtothey = x^y
naturallogx = log(x)
logbase10x = log10(x)
logbase2x = log2(x)
logbasearbx = log(x, base=42)
```

Note: The first five functions operate on a column-wise basis.

3.2.2   Trigonometric functions

```
sinpi = sin(pi)
cos0 = cos(0)
tanval = tan(pi/4)
acosx = acos(x)
asinx = asin(x)
atanx = atan(x)
atanxy = atan2(x, y)
```

3.2.3   Special functions

```
betaxy = beta(x, y)
gammax = gamma(x)
factorialn = factorial(n)
nchooser = choose(n, r)

library(gtools)
nchooser = length(combinations(n, r)[,1])
npermr = length(permutations(n, r)[,1])
```

Note: The `combinations()` and `permutations()` functions return a list of possible combinations and permutations.

3.2.4   Integer functions

See also 1.2.2 (rounding and number of digits to display).

```
nextintx = ceiling(x)
justintx = floor(x)
round2dec = round(x, 2)
roundint = round(x)
keep4sig = signif(x, 4)
movetozero = trunc(x)
```

Note: The second parameter of the `round()` function determines how many decimal places to round. The value of `movetozero` is the same as `justint` if $x > 0$ or `nextint` if $x < 0$.

3.2.5   Comparisons of floating-point variables

Because certain floating-point values of variables do not have exact decimal equivalents, there may be some error in how they are represented on a computer. For example, if the true value of a particular variable is $1/7$, the approximate decimal is 0.1428571428571428. For some operations (for example, tests of equality), this approximation can be problematic.

```
> all.equal(0.1428571, 1/7)
[1] "Mean relative difference: 3.000000900364093e-07"
> all.equal(0.1428571, 1/7, tolerance=0.0001)
[1] TRUE
```

Note: The tolerance option for the `all.equal()` function determines how many decimal places to use in the comparison of the vectors or scalars (the default tolerance is set to the underlying lower-level machine precision).

3.2.6   Complex numbers

Support for operations on complex numbers is available.

```
> (0+1i)^2    # i-squared
[1] -1+0i
```

Note: The above expression is equivalent to $i^2 = -1$. Additional support is available through the `complex()` function and related routines (see also `Re()` and `Im()`).

3.2.7   Derivatives

Rudimentary support for finding derivatives is available. These functions are particularly useful for high-dimensional optimization problems (see 3.2.9).

```
library(mosaic)
D(x^2 ~ x)
```

Note: The `D()` function within the `mosaic` package returns a function that can be evaluated or plotted using `plotFun()`. Second (or higher order) derivatives can be found by repeatedly applying the D function with respect to X. This function (as well as `deriv()`) is useful in numerical optimization (see the `nlm()`, `optim()` and `optimize()` functions).

3.2.8   Integration

Example: 10.1.6

Rudimentary support for calculus, including the evaluation of integrals, is available.

```
library(mosaic)
antiD(2*x ~ x)
```

Note: See also `integrate()`.

3.2.9 Optimization problems

R can be used to solve optimization (maximization) problems. As an extremely simple example, consider maximizing the area of a rectangle with perimeter equal to 20. Each of the sides can be represented by x and 10-x, with the area of the rectangle equal to $x * (10 - x)$.

```
f = function(x) { return(x*(10-x)) }
optimize(f, interval=c(0, 10), maximum=TRUE)
```

Note: Other optimization functions available within R include `nlm()`, `uniroot()`, `optim()`, and `constrOptim()` (see the CRAN task view on optimization and mathematical programming).

3.3 Matrix operations

Matrix operations are often needed in statistical analysis. Matrices can be created using the `matrix()` function (see A.4.5): matrix operations are then immediately available. In addition to the routines described below, the `Matrix` package in R is particularly useful for manipulation of large as well as sparse matrices.

3.3.1 Create matrix from vector

In this entry, we demonstrate creating a 2×2 matrix consisting of the first four nonzero integers:

$$A = \begin{pmatrix} 1 & 2 \\ 3 & 4 \end{pmatrix}.$$

```
A = matrix(c(1, 2, 3, 4), nrow=2, ncol=2, byrow=TRUE)
```

3.3.2 Combine vectors or matrices

We demonstrate creating a matrix from a set of conformable column vectors or smaller datasets.

```
A = cbind(x1, ..., xk)
A2 = c(x1, ..., xk)
```

Note: A is a matrix with columns made up of all the columns of $x_1, \ldots, x_k$. A2 is a vector of all of the elements in x_1, followed by all of the elements of x_2, etc. The `rbind()` function can be used to combine the matrices as rows instead of columns, making a $k \times n$ matrix (see also the `unstack()` and `stack()` commands).

3.3.3 Matrix addition

```
A = matrix(c(1, 2, 3, 4), nrow=2, ncol=2, byrow=TRUE)
B = A + A
```

3.3.4   Transpose matrix

```
A = matrix(c(1, 2, 3, 4), nrow=2, ncol=2, byrow=TRUE)
transA = t(A)
```

Note: If a dataframe is transposed in this manner, it will be converted to a matrix (which forces a single class for the objects within it).

3.3.5   Find the dimension of a matrix or dataset

```
A = matrix(c(1, 2, 3, 4), nrow=2, ncol=2, byrow=TRUE)
dim(A)
```

Note: The `dim()` function returns the dimension (number of rows and columns) for both matrices and dataframes, but does not work for vectors. The `length()` function returns the length of a vector or the number of elements in a matrix.

3.3.6   Matrix multiplication

```
A = matrix(c(1, 2, 3, 4), nrow=2, ncol=2, byrow=TRUE)
Asquared = A %*% A
```

3.3.7   Finding the inverse of a matrix

```
A = matrix(c(1, 2, 3, 4), nrow=2, ncol=2, byrow=TRUE)
Ainv = solve(A)
```

3.3.8   Component-wise multiplication

Unlike the matrix multiplication in 3.3.6, the result of this operation is scalar multiplication of each element in the matrix. For example, the component-wise multiplication of

$$\begin{pmatrix} 1 & 2 \\ 3 & 4 \end{pmatrix} \text{ with itself yields } \begin{pmatrix} 1 & 4 \\ 9 & 16 \end{pmatrix}.$$

```
A = matrix(c(1, 2, 3, 4), nrow=2, ncol=2, byrow=TRUE)
newmat = A * A
```

3.3.9   Create a submatrix

```
A = matrix(1:12, nrow=3, ncol=4, byrow=TRUE)
Asub = A[2:3, 3:4]
```

3.3.10   Create a diagonal matrix

```
A = matrix(c(1, 2, 3, 4), nrow=2, ncol=2, byrow=TRUE)
diagMat = diag(c(1, 4))     # argument is a vector
diagMat = diag(diag(A))     # A is a matrix
```

Note: For a vector argument, the `diag()` function generates a matrix with the vector values as the diagonals and all off-diagonals 0. For matrix A, the `diag()` function creates a vector of the diagonal elements (see 3.3.11); a diagonal matrix with these diagonal entries, but all off-diagonals set to 0, can be created by running the `diag()` with this vector as an argument.

3.3.11 Create a vector of diagonal elements

```
A = matrix(c(1, 2, 3, 4), nrow=2, ncol=2, byrow=TRUE)
diagVals = diag(A)
```

3.3.12 Create a vector from a matrix

Note: This makes a row vector from all the values in the matrix.

```
A = matrix(c(1, 2, 3, 4), nrow=2, ncol=2, byrow=TRUE)
newvec = c(A)
```

3.3.13 Calculate the determinant

```
A = matrix(c(1, 2, 3, 4), nrow=2, ncol=2, byrow=TRUE)
det(A)
```

3.3.14 Find eigenvalues and eigenvectors

```
A = matrix(c(1, 2, 3, 4), nrow=2, ncol=2, byrow=TRUE)
Aev = eigen(A)
Aeval = Aev$values
Aevec = Aev$vectors
```

Note: The `eigen()` function in R returns a list consisting of the eigenvalues and eigenvectors, respectively, of the matrix given as the argument.

3.3.15 Find the singular value decomposition

The singular value decomposition of a matrix A is given by $A = U * \mathrm{diag}(Q) * V^T$, where $U^T U = V^T V = V V^T = I$ and Q contains the singular values of A.

```
A = matrix(c(1, 2, 3, 4), nrow=2, ncol=2, byrow=TRUE)
svdres = svd(A)
U = svdres$u
Q = svdres$d
V = svdres$v
```

Note: The `svd()` function returns a list with components corresponding to a vector of singular values, a matrix with columns corresponding to the left singular values, and a matrix with columns containing the right singular values.

3.4   Examples

To help illustrate the tools presented in this chapter, we apply some of the entries in examples. The R code can be downloaded from http://www.amherst.edu/~nhorton/r2/ examples.

3.4.1   Probability distributions

To demonstrate more tools, we leave the HELP dataset and show examples of how data can be generated. We will generate values (3.1.6) from the normal and t distribution densities.

```
> x = seq(from=-4, to=4.2, length=100)
> normval = dnorm(x, 0, 1)
> dfval = 1
> tval = dt(x, df=dfval)
```

Figure 3.1 displays a plot of these distributions.

The `xpnorm()` function within the `mosaic` package may be useful for teaching purposes, to display information about the normal density function (see Figure 3.2).

Other distributions (e.g., the exponential) can easily be displayed using the `plotDist()` function within the `mosaic` package.

```
> plot(x, normval, type="n", ylab="f(x)", las=1)
> lines(x, normval, lty=1, lwd=2)
> lines(x, tval, lty=2, lwd=2)
> legend(1.1, .395, lty=1:2, lwd=2,
      legend=c(expression(N(mu == 0,sigma == 1)),
      paste("t with ", dfval," df", sep="")))
```

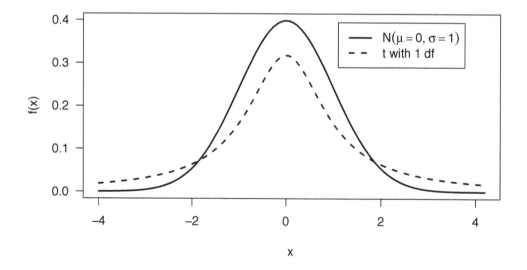

Figure 3.1: Comparison of standard normal and t distribution with 1 df

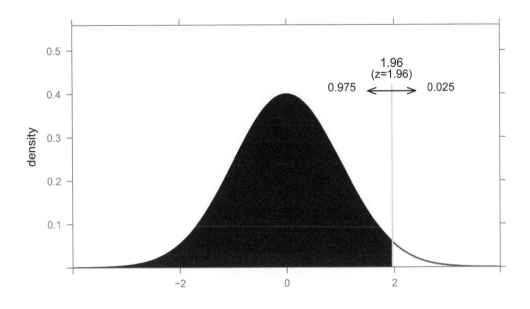

Figure 3.2: Descriptive plot of the normal distribution

```
> library(mosaic)
> xpnorm(1.96, mean=0, sd=1)

If X ~ N(0,1), then

P(X <= 1.96) = P(Z <= 1.96) = 0.975
P(X >  1.96) = P(Z >  1.96) = 0.025


[1] 0.975
```

Chapter 4

Programming and operating system interface

This chapter reviews programming functions as well as interactions with the underlying operating system.

4.1 Control flow, programming, and data generation

4.1.1 Looping

Example: 11.2

```
x = numeric(k)   # create placeholder
for (i in 1:length(x)) {
    x[i] = rnorm(1) # this is slow and inefficient!
}
```

or (preferably)

```
x = rnorm(k)  # this is far better
```

Note: Most tasks in R that could be written as a loop are often dramatically faster if they are encoded as a vector operation (as in the second and preferred option above). Examples of situations where loops are particularly useful can be found in 11.1.2 and 11.2. The `along.with` option for `seq()` and the `seq_along()` function can also be helpful.

More information on control structures for looping and conditional processing such as `while` and `repeat` can be found in `help(Control)`.

4.1.2 Conditional execution

Examples: 6.6.6 and 8.7.7

```
if (expression1) { expression2 }
```

or

```
if (expression1) { expression2 } else { expression3 }
```

or

```
ifelse(expression, x, y)
```

Note: The `if` statement, with or without `else`, tests a single logical statement; it is not an elementwise (vector) function. If `expression1` evaluates to TRUE, then `expression2` is evaluated. The `ifelse()` function operates on vectors and evaluates the expression given as `expression` and returns x if it is TRUE and y otherwise (see comparisons, A.4.2). An expression can include multi-command blocks of code (in brackets). The `switch()` function may also be useful for more complicated tasks.

4.1.3 Sequence of values or patterns

Example: 10.1.3

It is often useful to generate a variable consisting of a sequence of values (e.g., the integers from 1 to 100) or a pattern of values (1 1 1 2 2 2 3 3 3). This might be needed to generate a variable consisting of a set of repeated values for use in a simulation or graphical display.

As an example, we demonstrate generating data from a linear regression model of the form:

$$E[Y|X_1, X_2] = \beta_0 + \beta_1 X_1 + \beta_2 X_2, \ Var(Y|X) = 9, \ Corr(X_1, X_2) = 0.$$

```
# generate
seq(from=i1, to=i2, length.out=nvals)
seq(from=i1, to=i2, by=1)
seq(i1, i2)
i1:i2

rep(value, times=nvals)
```
or
```
rep(value, each=nvals)
```

Note: The `seq` function creates a vector of length `val` if the `length.out` option is specified. If the `by` option is included, the length is approximately `(i2-i1)/byval`. The `i1:i2` operator is equivalent to `seq(from=i1, to=i2, by=1)`. The `rep` function creates a vector of length `nvals` with all values equal to `value`, which can be a scalar, vector, or list. The `each` option repeats each element of `value` `nvals` times. The default is `times`.

The following code implements the model described above for $n = 200$.

```
> n = 200
> x1 = rep(c(0,1), each=n/2)      # x1 resembles 0 0 0 ... 1 1 1
> x2 = rep(c(0,1), n/2)           # x2 resembles 0 1 0 1 ... 0 1
> beta0 = -1; beta1 = 1.5; beta2 = .5;
> rmse = 3
> table(x1, x2)
   x2
x1   0  1
  0 50 50
  1 50 50
> y = beta0 + beta1*x1 + beta2*x2 + rnorm(n, mean=0, sd=rmse)
> lm(y ~ x1 + x2)
```

4.1.4 Perform an action repeatedly over a set of variables

It is often necessary to perform a given function for a series of variables. Here, the square of each of a list of variables is calculated as an example.

```
l1 = c("x1", "x2", ..., "xk")
l2 = c("z1", "z2", ..., "zk")
for (i in 1:length(l1)) {
   assign(l2[i], eval(as.name(l1[i]))^2)
}
```

Note: It is not straightforward to refer to objects without evaluating those objects. Assignments to R objects given symbolically can be made using the `assign()` function. Here, a non-obvious use of the `eval()` function is used to evaluate an expression after the string value in `l1` is coerced to be a symbol. This allows the values of the character vectors `l1` and `l2` to be evaluated (see `help(assign)`, `eval()`, and `substitute()`).

4.1.5 Grid of values

Example: 12.8

It may be useful to generate all combinations of two or more vectors.

```
> expand.grid(x1=1:3, x2=c("M", "F"))
  x1 x2
1  1  M
2  2  M
3  3  M
4  1  F
5  2  F
6  3  F
```

Note: The `expand.grid()` function takes two or more vectors or factors and returns a dataframe. The first factor varies fastest. The resulting object is a matrix.

4.1.6 Debugging

```
browser()   # create a breakpoint
debug(function)  # enter the debugger when function called
```

Note: When a function flagged for debugging is called, the function can be executed one statement at a time. At the prompt, commands can be entered (n for next, c for continue, `where` for traceback, Q for quit) or expressions can be evaluated (see `browser()` and `trace()`). A debugging environment is available within RStudio. The debugger may be invoked by setting a breakpoint by clicking to the left of the line number in a script, or pressing Shift+F9. Profiling of the execution of expressions can be undertaken using the `Rprof()` function (see also `summaryRprof()` and `tracemem()`). RStudio provides a series of additional debugging tools.

4.1.7 Error recovery

```
try(expression, silent=FALSE)
stopifnot(expr1, ..., exprk)
```

Note: The `try()` function runs the given `expression` and traps any errors that may arise (displaying them on the standard error output device). The function `geterrmessage()` can be used to display any errors. The `stopifnot()` function runs the given expressions and returns an error message if all are not true (see `stop()` and `message()`).

4.2   Functions

A strength of R is its extensibility. In this section, we provide an introduction to defining and calling functions.

A new function is defined by the syntax `function(arglist)` body. The body is made up of a series of commands (or expressions), typically separated by line breaks and enclosed in curly braces. Here, we create a function to calculate the estimated confidence interval (CI) for a mean, as in 5.1.7.

```
# calculate a t confidence interval for a mean
ci.calc = function(x, ci.conf=.95) {
    sampsize = length(x)
    tcrit = qt(1-((1-ci.conf)/2), sampsize - 1)
    mymean = mean(x)
    mysd = sd(x)
    return(list(civals=c(mymean-tcrit*mysd/sqrt(sampsize),
        mymean+tcrit*mysd/sqrt(sampsize)), ci.conf=ci.conf))
}
```

Here, the appropriate quantile of the T distribution is calculated using the `qt()` function, and the appropriate confidence interval is calculated and returned as a list. The function is stored in the object `ci.calc`, which can then be used like any other function. For example, `ci.calc(x1)` will print the CI and confidence level for the object `x1`. We also demonstrate the syntax for providing a default value, so that the confidence level in the preceding example is 0.95. User-written functions nest just as pre-existing functions do: `ci.conf(rnorm(100), 0.9)` will return the CI and report that the confidence limit is 0.9 for 100 normal random variates. In this example, we explicitly `return()` a list of return values. If no return statement is provided, the results of the last expression evaluation are returned.

```
> ci.calc(x)
$civals
[1]   0.624 12.043
$ci.conf
[1] 0.95
```

If only the lower confidence interval is needed, this can be saved as an object.

```
> lci = ci.calc(x)$civals[1]
> lci
[1] 0.624
```

The default confidence level is 95%; this can be changed by specifying a different value.

```
> ci.calc(x, ci.conf=.90)
$civals
[1]   1.799 10.867

$ci.conf
[1] 0.9
```

This is equivalent to running `ci.calc(x, .90)`, since `ci.conf` is the second argument to the function.

Other examples of defined functions can be found in 2.2.18 and 5.7.4.

4.3 Interactions with the operating system

4.3.1 Timing commands

See also 2.4.6 (time variables).

```
system.time(expression)
```

Note: The `expression` (e.g., call to any user-or system-defined function, see A.4.1) given as an argument to the `system.time()` function is evaluated, and the user, system, and total (elapsed) time are returned (see `proc.time()`).

4.3.2 Suspend execution for a time interval

```
Sys.sleep(numseconds)
```

Note: The command `Sys.sleep()` will pause execution for `numseconds`, with minimal system impact.

4.3.3 Execute a command in the operating system

```
system("ls")
```

Note: The command `ls` lists the files in the current working directory (see 4.3.7 to capture this information). When running under Windows, the `shell()` command can be used to start a command window.

4.3.4 Command history

```
savehistory()
loadhistory()
history()
```

Note: The command `savehistory()` saves the history of recent commands, which can be re-loaded using `loadhistory()` or displayed using `history()`. The `timestamp()` function can be used to add a date and time stamp to the history.

4.3.5 Find working directory

```
getwd()
```

Note: The command `getwd()` displays the current working directory.

4.3.6 Change working directory

```
setwd("dir_location")
```

Note: The command `setwd()` changes the current working directory to the (absolute or relative) pathname given as an argument (see `file.choose()`). This can also be done interactively under Windows and Mac OS X by selecting the `Change Working Directory` option under the `Misc` menu, or similar options on the `Session` menu in RStudio.

4.3.7 List and access files

```
list.files()
```

Note: The `list.files()` command returns a character vector of filenames in the current directory (by default). Recursive listings are also supported. The function `file.choose()` provides an interactive file browser and can be given as an argument to functions such as `read.table()` (1.1.2) or `read.csv()` (1.1.4). Related file operation functions include `file.access()`, `file.exists()`, `file.info()`, and `unlink()` (see `help(files)` and `Sys.glob()` for wildcard expansion).

4.3.8 Create temporary file

```
uniquefile = tempfile()
cat(x, "\n", file=uniquefile)
```

Note: The filenames returned by `tempfile()` are likely to be unique among calls in an R session (and guaranteed not to be currently in use).

4.3.9 Redirect output

```
capture.output(Sys.time(), file="filename")
```
or
```
sink(file="filename")
Sys.time()
sink()
```

Note: The result of the first argument to `capture.output()` is saved as text in the filename. This can be particularly useful for post-processing functions with long output. The `sink()` function diverts all output and/or messages to a connection, until the `sink()` function is run again. In this example, the output of `Sys.time()` is saved to a file.

Chapter 5

Common statistical procedures

This chapter describes how to generate univariate summary statistics (such as means, variances, and quantiles) for continuous variables, display and analyze frequency tables and cross-tabulations of categorical variables, and carry out a variety of one- and two-sample procedures.

5.1 Summary statistics

5.1.1 Means and other summary statistics

Example: 5.7.1

```
mean(x)
```
or
```
library(mosaic)
favstats(x, data=ds)
```

Note: The `mean()` function accepts a numeric vector or a numeric dataframe as arguments (date objects are also supported). Similar functions include `median()` (see 5.1.5 for more quantiles), `var()`, `sd()`, `min()`, `max()`, `sum()`, `prod()`, and `range()` (note that the latter returns a vector containing the minimum and maximum value). The `rowMeans()` and `rowSums()` functions (and their equivalents for columns) can be helpful for some calculations. The `favstats()` function in the `mosaic` package provides a concise summary of the distribution of a variable (including the number of observations and missing values). Discussion of how to calculate summary statistics by group can be found in 11.1.

5.1.2 Weighted means and other statistics

```
library(Hmisc)
wtd.mean(x, weights)
```

Note: The `wtd.mean()` function in the `Hmisc` package calculates weighted means. Other related functions include `wtd.var()`, `wtd.quantile()`, and `wtd.rank()`.

5.1.3　Other moments

Example: 5.7.1

While skewness and kurtosis are less commonly reported than the mean and standard deviation, they can be useful at times. Skewness is defined as the third moment around the mean and characterizes whether the distribution is symmetric (skewness=0). Kurtosis is a function of the fourth central moment. It characterizes peakedness, where the normal distribution has a value of 3 and smaller values correspond to a more rounded peak and shorter, thinner tails.

```
library(moments)
skewness(x)
kurtosis(x)
```

Note: The `moments` package facilitates the calculation of skewness and kurtosis within R as well as higher-order moments (see `all.moments()`).

5.1.4　Trimmed mean

```
mean(x, trim=frac)
```

Note: The value `frac` can take on range 0 to 0.5 and specifies the fraction of observations to be trimmed from each end of `x` before the mean is computed (`frac=0.5` yields the median).

5.1.5　Quantiles

Example: 5.7.1

```
quantile(x, c(.025, .975))
quantile(x, seq(from=.95, to=.975, by=.0125))
```

Note: Details regarding the calculation of quantiles in `quantile()` can be found using `help(quantile)`. The `ntiles()` function in the `mosaic` package can facilitate the creation of groups of roughly equal sizes.

5.1.6　Centering, normalizing, and scaling

```
scale(x)
```
or
```
(x-mean(x))/sd(x)
```
Note: The default behavior of `scale()` is to create a Z-score transformation. The `scale()` function can operate on matrices and dataframes, and allows the specification of a vector of the scaling parameters for both center and scale (see `sweep()`, a more general function).

5.1.7　Mean and 95% confidence interval

```
tcrit = qt(.975, df=length(x)-1)
mean(x) + c(-tcrit, tcrit)*sd(x)/sqrt(length(x))
```
or
```
t.test(x)$conf.int
```

Note: While the appropriate 95% confidence interval can be generated in terms of the mean and standard deviation, it is more straightforward to use the `t.test()` function to calculate the relevant quantities.

5.1.8 Proportion and 95% confidence interval

Example: 11.2

```
binom.test(sum(x), length(x))
prop.test(sum(x), length(x))
```

Note: The `binom.test()` function calculates an exact Clopper–Pearson confidence interval based on the F distribution [25] using the first argument as the number of successes and the second argument as the number of trials, while `prop.test()` calculates an approximate confidence interval by inverting the score test. Both allow specification of the probability under the null hypothesis. The `conf.level` option can be used to change the default confidence level.

5.1.9 Maximum likelihood estimation of parameters

Example: 5.7.1

See also 3.1.1 (probability density functions).

```
library(MASS)
fitdistr(x, "densityfunction")
```

Note: Options for `densityfunction` include `beta`, `cauchy`, `chi-squared`, `exponential`, `f`, `gamma`, `geometric`, `log-normal`, `lognormal`, `logistic`, `negative binomial`, `normal`, `Poisson`, `t`, and `weibull`.

5.2 Bivariate statistics

5.2.1 Epidemiologic statistics

Example: 5.7.3

```
sum(x==0&y==0)*sum(x==1&y==1)/(sum(x==0&y==1)*sum(x==1&y==0))
```
or
```
tab1 = table(x, y)
tab1[1,1]*tab1[2,2]/(tab1[1,2]*tab1[2,1])
```
or
```
glm1 = glm(y ~ x, family=binomial)
exp(glm1$coef[2])
```
or
```
library(epitools)
oddsratio.fisher(x, y)
oddsratio.wald(x, y)
riskratio(x, y)
riskratio.wald(x, y)
```

Note: The `epitab()` function in the `epitools` package provides a general interface to many epidemiologic statistics, while `expand.table()` can be used to create individual level data from a table of counts (see generalized linear models, 7.1).

5.2.2   Test characteristics

The sensitivity of a test is defined as the probability that someone with the disease ($D = 1$) tests positive ($T = 1$), while the specificity is the probability that someone without the disease ($D = 0$) tests negative ($T = 0$). For a dichotomous screening measure, the sensitivity and specificity can be defined as $P(D = 1, T = 1)/P(D = 1)$ and $P(D = 0, T = 0)/P(D = 0)$, respectively (see also receiver operating characteristic curves, 8.5.7).

```
sens = sum(D==1&T==1)/sum(D==1)
spec = sum(D==0&T==0)/sum(D==0)
```

Note: Sensitivity and specificity for an outcome D can be calculated for each value of a continuous measure T using the following code.

```
library(ROCR)
pred = prediction(T, D)
diagobj = performance(pred, "sens", "spec")
spec = slot(diagobj, "y.values")[[1]]
sens = slot(diagobj, "x.values")[[1]]
cut = slot(diagobj, "alpha.values")[[1]]
diagmat = cbind(cut, sens, spec)
head(diagmat, 10)
```

Note: The ROCR package facilitates the calculation of test characteristics, including sensitivity and specificity. The prediction() function takes as arguments the continuous measure and outcome. The returned object can be used to calculate quantities of interest (see help(performance) for a comprehensive list). The slot() function is used to return the desired sensitivity and specificity values for each cut score, where [[1]] denotes the first element of the returned list (see help(list) and help(Extract)).

5.2.3   Correlation

Examples: 5.7.2 and 8.7.7

```
pearsoncorr = cor(x, y)
spearmancorr = cor(x, y, method="spearman")
kendalltau = cor(x, y, method="kendall")
```
or
```
cormat = cor(cbind(x1, ..., xk))
```

Note: Specifying method="spearman" or method="kendall" as an option to cor() generates the Spearman or Kendall correlation coefficients, respectively. A matrix of variables (created with cbind()) can be used to generate the correlation between a set of variables. The use option for cor() specifies how missing values are handled (either "all.obs", "complete.obs", or "pairwise.complete.obs"). The cor.test() function can carry out a test (or calculate the confidence interval) for a correlation.

5.2.4   Kappa (agreement)

```
library(irr)
kappa2(data.frame(x, y))
```

Note: The kappa2() function takes a dataframe (see A.4.6) as an argument. Weights can be specified as an option.

5.3 Contingency tables

5.3.1 Display cross-classification table

Example: 5.7.3

Contingency tables show the group membership across categorical (grouping) variables. They are also known as cross-classification tables, cross-tabulations, and two-way tables.

```
library(gmodels)
CrossTable(x, y)
```
or
```
mytab = table(y, x)
addmargins(mytab)
```
or
```
library(mosaic)
tally(~ y + x, margins=TRUE, data=ds)
```
or
```
library(prettyR)
xtab(y ~ x, data=ds)
```

Note: The `CrossTable()` function in the `gmodels` package provides a flexible means to generate crosstabs. It supports the `missing.include` option to add a category for missing values, unused factor levels, as well as emulation of SPSS or SAS output, with cell, row, and/or column percentages. For the `table()` function, the `exclude=NULL` option includes categories for missing values. The `addmargins()` function adds (by default) the row and column totals to a table. The `colSums()`, `colMeans()` functions (and their equivalents for rows) can be used to efficiently calculate sums and means for numeric vectors. The `tally()` function in the `mosaic` package supports a modeling language for categorical tables, including a | operator to stratify by a third variable. Options for the `tally()` function include `format=` (`percent`, `proportion`, or `count`) and `margins=`. Additional options for table display are provided in the `prettyR` package `xtab()` function.

5.3.2 Displaying missing value categories in a table

It can be useful to display tables including missing values as a separate category (see 11.4.4.1).

```
table(x1, x2, useNA="ifany")
```

5.3.3 Pearson chi-square statistic

Example: 5.7.3

```
chisq.test(x, y)
```
or
```
chisq.test(ymat)
```

Note: The `chisq.test()` command can accept either two class vectors or a table of counts. By default, a continuity correction is used (the option `correct=FALSE` turns this off). A version with more verbose output (e.g., expected cell counts) can be found in the `xchisq.test()` function in the `mosaic` package.

5.3.4 Cochran–Mantel–Haenszel test

The Cochran–Mantel–Haenszel test gives an assessment of the relationship between X_1 and X_2, stratified by (or controlling for) X_3. The analysis provides a way to adjust for the possible confounding effects of X_3 without having to estimate parameters for them.

```
mantelhaen.test(x1, x2, x3)
```

5.3.5　Cramér's V

Cramér's V (or phi coefficient) is a measure of association for nominal variables.

```
library(vcd)
assocstats(table(x, y))
```

5.3.6　Fisher's exact test

Example: 5.7.3

```
fisher.test(y, x)
```
or
```
fisher.test(ymat)
```

Note: The `fisher.test()` command can accept either two class vectors or a table of counts (here denoted by `ymat`). For tables with many rows and/or columns, p-values can be computed using Monte Carlo simulation using the `simulate.p.value` option. The Monte Carlo p-value can be considerably less compute-intensive for large sample sizes.

5.3.7　McNemar's test

McNemar's test tests the null hypothesis that the proportions are equal across matched pairs, for example, when two raters assess a population.

```
mcnemar.test(y, x)
```

Note: The `mcnemar.test()` command can accept either two class vectors or a matrix with counts.

5.4　Tests for continuous variables

5.4.1　Tests for normality

```
shapiro.test(x)
```

Note: The `nortest` package includes a number of additional tests of normality.

5.4.2　Student's t-test

Example: 5.7.4

```
t.test(y1, y2)
```
or
```
t.test(y ~ x, data=ds)
```

Note: The first example for the `t.test()` command displays how it can take two vectors (`y1` and `y2`) as arguments to compare, or in the latter example, a single vector corresponding to the outcome (`y`), with another vector indicating group membership (`x`) using a formula interface (see A.4.7 and 6.1.1). By default, the two-sample t-test uses an unequal variance assumption. The option `var.equal=TRUE` can be added to specify an equal variance assumption. The command `var.test()` can be used to formally test equality of variances.

5.4.3 Test for equal variances

The assumption of equal variances among the groups in analysis of variance and the two-sample t-test can be assessed via Levene's test.

```
var.test(y, x)
```

or

```
library(lawstat)
levene.test(y, x, location="mean")
bartlett.test(y ~ x)
```

Note: Other options to assess equal variance include Bartlett's test, the Brown and Forsythe version of Levene's test, and O'Brien's test, which is effectively a modification of Levene's test. Options to the `levene.test()` function provide other variants.

5.4.4 Nonparametric tests

Example: 5.7.4

```
wilcox.test(y1, y2)
ks.test(y1, y2)

library(coin)
median_test(y ~ x)
```

Note: By default, the `wilcox.test()` function uses a continuity correction in the normal approximation for the p-value. The `ks.test()` function does not calculate an exact p-value when there are ties. The median test shown will generate an exact p-value with the `distribution="exact"` option.

5.4.5 Permutation test

Example: 5.7.4

```
library(coin)
oneway_test(y ~ as.factor(x), distribution=approximate(B=bnum))
```

or

```
library(mosaic)
obs = t.test(y ~ x)$statistic
res = do(10000) * t.test(y ~ shuffle(x))$statistic
tally(~ t > abs(obs), data=res)
```

Note: The `oneway_test` function in the `coin` package implements a variety of permutation-based tests (see the `exactRankTests` package). The `distribution=approximate` syntax generates an empirical p-value (asymptotically equivalent to the exact p-value) based on `bnum` Monte Carlo replicates. The `do()` function along with the `shuffle()` functions in the `mosaic` package can also be used to undertake permutation tests (see the package's resampling vignette at CRAN for details).

5.4.6   Logrank test

Example: 5.7.5

See also 8.5.11 (Kaplan–Meier plot) and 7.5.1 (Cox proportional hazards model).

```
library(survival)
survdiff(Surv(timevar, cens) ~ x)
```

Note: Other tests within the G-rho family of Fleming and Harrington [41] are supported by specifying the `rho` option.

5.5   Analytic power and sample size calculations

Many simple settings lend themselves to analytic power calculations, where closed form solutions are available. Other situations may require an empirical calculation, where repeated simulation is undertaken (see 11.2).

It is straightforward to find power or sample size (given a desired power) for two-sample comparisons of either continuous or categorical outcomes. We show simple examples for comparing means and proportions in two groups and supply additional information on analytic power calculation available for more complex methods.

```
# find sample size for two-sample t-test
power.t.test(delta=0.5, power=0.9)
```

```
# find power for two-sample t-test
power.t.test(delta=0.5, n=100)
```

The latter call generates the following output.

```
     Two-sample t test power calculation
              n = 100
          delta = 0.5
             sd = 1
      sig.level = 0.05
          power = 0.9404272
    alternative = two.sided
 NOTE: n is number in *each* group
```

```
# find sample size for two-sample test of proportions
power.prop.test(p1=.1, p2=.2, power=.9)
```

```
# find power for two-sample test of proportions
power.prop.test(p1=.1, p2=.2, n=100)
```

Note: The `power.t.test()` function requires exactly four of the five arguments (sample size in each group, power, difference between groups, standard deviation, and significance level) to be specified. Default values exist for `sd=1` and `sig.level=0.05`. Other power calculation functions can be found in the `pwr` package.

5.6 Further resources

A comprehensive introduction to using R to fit common statistical models can be found in [181]. A readable introduction to permutation-based inference can be found in [104]. A vignette on resampling-based inference using R can be found at `http://cran.r-project.org/web/packages/mosaic/vignettes/V5Resample.pdf`. Collett [26] provides an accessible introduction to survival analysis.

5.7 Examples

To help illustrate the tools presented in this chapter, we apply many of the entries to the HELP data. The code can be downloaded from `http://www.amherst.edu/~nhorton/r2/examples`.

5.7.1 Summary statistics and exploratory data analysis

We begin by reading the dataset.

```
> options(digits=3)
> options(width=72)   # narrows output to stay in the grey box
> ds = read.csv("http://www.amherst.edu/~nhorton/r2/datasets/help.csv")
```

A first step is to examine some univariate statistics (5.1.1) for the baseline CESD (Center for Epidemiologic Statistics) measure of depressive symptoms score. We can use functions that produce a set of statistics, such as `favstats()`, from the `mosaic` package, and request them singly.

```
> with(ds, mean(cesd))
[1] 32.8
```

```
> with(ds, median(cesd))
[1] 34
> with(ds, range(cesd))
[1]  1 60
> with(ds, sd(cesd))
[1] 12.5
> with(ds, var(cesd))
[1] 157
> library(mosaic)
> favstats(~ cesd, data=ds)
 min Q1 median Q3 max mean   sd   n missing
   1 25     34 41  60 32.8 12.5 453       0
```

```
> library(moments)
> with(ds, skewness(cesd))
[1] -0.26
> with(ds, kurtosis(cesd))
[1] 2.55
```

We can also generate desired quantiles. Here, we find the deciles (5.1.5).

```
> with(ds, quantile(cesd, seq(from=0, to=1, length=11)))
  0%  10%  20%  30%  40%  50%  60%  70%  80%  90% 100%
 1.0 15.2 22.0 27.0 30.0 34.0 37.0 40.0 44.0 49.0 60.0
```

Graphics can allow us to easily review the whole distribution of the data. Here, we generate a histogram (8.1.4) of CESD, overlaid with its empirical PDF (8.1.5) and the closest-fitting normal distribution (see Figure 5.1).

```
> with(ds, hist(cesd, main="", freq=FALSE))
> with(ds, lines(density(cesd), main="CESD", lty=2, lwd=2))
> xvals = with(ds, seq(from=min(cesd), to=max(cesd), length=100))
> with(ds, lines(xvals, dnorm(xvals, mean(cesd), sd(cesd)), lwd=2))
```

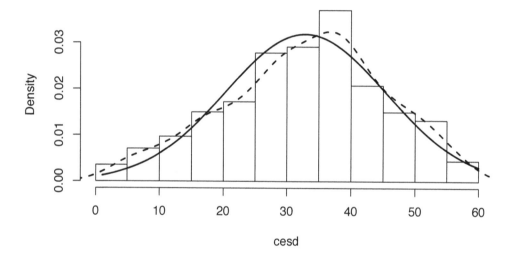

Figure 5.1: Density plot of depressive symptom scores (CESD) plus superimposed histogram and normal distribution

5.7.2   Bivariate relationships

We can calculate the correlation (5.2.3) between CESD and MCS and PCS (mental and physical component scores). First, we show the default correlation matrix.

```
> cormat = cor(with(ds, cbind(cesd, mcs, pcs)))
> cormat
        cesd    mcs    pcs
cesd   1.000 -0.682 -0.293
mcs   -0.682  1.000  0.110
pcs   -0.293  0.110  1.000
```

To save space, we can just print a subset of the correlations.

```
> cormat[c(2, 3), 1]
   mcs    pcs
-0.682 -0.293
```

Figure 5.2 displays a scatterplot (8.3.1) of CESD and MCS, for the female subjects. The plotting character (9.1.2) is the initial letter of the primary substance (alcohol, cocaine, or heroin). A rug plot (9.1.8) is added to help demonstrate the marginal distributions.

```
> with(ds, plot(cesd[female==1], mcs[female==1], xlab="CESD", ylab="MCS",
    type="n", bty="n"))
> with(ds, text(cesd[female==1&substance=="alcohol"],
    mcs[female==1&substance=="alcohol"],"A"))
> with(ds, text(cesd[female==1&substance=="cocaine"],
    mcs[female==1&substance=="cocaine"],"C"))
> with(ds, text(cesd[female==1&substance=="heroin"],
    mcs[female==1&substance=="heroin"],"H"))
> with(ds, rug(jitter(mcs[female==1]), side=2))
> with(ds, rug(jitter(cesd[female==1]), side=3))
```

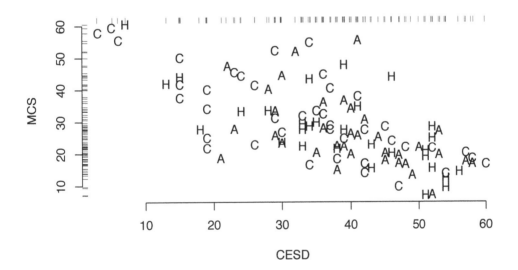

Figure 5.2: Scatterplot of CESD and MCS for women, with primary substance shown as the plot symbol

5.7.3 Contingency tables

Here we display the cross-classification (contingency) table (5.3.1) of homeless at baseline by gender, calculate the observed odds ratio (OR, 5.2.1), and assess association using the Pearson χ^2 test (5.3.3) and Fisher's exact test (5.3.6). The CrossTable() function from the gmodels package displays contingency tables, using the SPSS format.

```
> require(gmodels)
> with(ds, CrossTable(homeless, female, prop.chisq=FALSE, format="SPSS"))

   Cell Contents
|-----------------------|
|                 Count |
|           Row Percent |
|        Column Percent |
|         Total Percent |
|-----------------------|

Total Observations in Table:  453

             | female
   homeless  |        0  |         1  | Row Total |
-------------|-----------|------------|-----------|
          0  |      177  |        67  |      244  |
             |  72.541%  |   27.459%  |  53.863%  |
             |  51.156%  |   62.617%  |           |
             |  39.073%  |   14.790%  |           |
-------------|-----------|------------|-----------|
          1  |      169  |        40  |      209  |
             |  80.861%  |   19.139%  |  46.137%  |
             |  48.844%  |   37.383%  |           |
             |  37.307%  |    8.830%  |           |
-------------|-----------|------------|-----------|
Column Total |      346  |       107  |      453  |
             |  76.380%  |   23.620%  |           |
-------------|-----------|------------|-----------|
```

We can easily calculate the odds ratio directly.

```
> or = with(ds, (sum(homeless==0 & female==0)*
        sum(homeless==1 & female==1))/
       (sum(homeless==0 & female==1)*
        sum(homeless==1 & female==0)))
> or
[1] 0.625
```

We can also use the epitools package, which will generate confidence limits in addition to the odds ratio.

```
> library(epitools)
> oddsobject = with(ds, oddsratio.wald(homeless, female))
> oddsobject$measure
          odds ratio with 95% C.I.
Predictor estimate lower upper
        0    1.000    NA    NA
        1    0.625 0.401 0.975
```

```
> oddsobject$p.value
          two-sided
Predictor midp.exact fisher.exact chi.square
       0        NA           NA          NA
       1    0.0381       0.0456      0.0377
```

The χ^2 and Fisher's exact tests are fit in R using separate commands.

```
> chisqval = with(ds, chisq.test(homeless, female, correct=FALSE))
> chisqval

Pearson's Chi-squared test

data:  homeless and female
X-squared = 4.32, df = 1, p-value = 0.03767
```

```
> with(ds, fisher.test(homeless, female))

Fisher's Exact Test for Count Data

data:  homeless and female
p-value = 0.0456
alternative hypothesis: true odds ratio is not equal to 1
95 percent confidence interval:
 0.389 0.997
sample estimates:
odds ratio
    0.626
```

A graphical depiction of a table can be created (see Figure 5.3); this can helpful as part of automated report generation and reproducible analysis (see 11.3).

```
> library(gridExtra)
> mytab = tally(~ racegrp + substance, data=ds)
> plot.new()
> grid.table(mytab)
```

	alcohol	cocaine	heroin
black	55	125	31
hispanic	17	7	26
other	9	7	10
white	96	13	57

Figure 5.3: Graphical display of the table of substance by race/ethnicity

5.7.4   Two sample tests of continuous variables

We can assess gender differences in baseline age using a *t*-test (5.4.2) and nonparametric procedures.

```
> ttres = t.test(age ~ female, data=ds)
> print(ttres)


Welch Two Sample t-test

data:  age by female
t = -0.93, df = 180, p-value = 0.3537
alternative hypothesis: true difference in means is not equal to 0
95 percent confidence interval:
 -2.45  0.88
sample estimates:
mean in group 0 mean in group 1
          35.5            36.3
```

The **names()** function can be used to identify the objects returned by the **t.test()** function (not displayed).

A permutation test can be run and used to generate a Monte Carlo *p*-value (5.4.5).

```
> library(coin)
> oneway_test(age ~ as.factor(female),
     distribution=approximate(B=9999), data=ds)

Approximative 2-Sample Permutation Test

data:  age by as.factor(female) (0, 1)
Z = -0.919, p-value = 0.3493
alternative hypothesis: true mu is not equal to 0
```

```
> with(ds, wilcox.test(age ~ as.factor(female), correct=FALSE))

Wilcoxon rank sum test

data:  age by as.factor(female)
W = 17512, p-value = 0.3979
alternative hypothesis: true location shift is not equal to 0
```

```
> ksres = with(ds, ks.test(age[female==1], age[female==0]))
Warning:  p-value will be approximate in the presence of ties
> print(ksres)

Two-sample Kolmogorov-Smirnov test

data:  age[female == 1] and age[female == 0]
D = 0.063, p-value = 0.902
alternative hypothesis: two-sided
```

We can also plot estimated density functions (8.1.5) for age for both groups, and shade some areas (9.1.13) to emphasize how they overlap. We create a function (4.2) to automate this task.

```
> plotdens = function(x,y, mytitle, mylab) {
    densx = density(x)
    densy = density(y)
    plot(densx, main=mytitle, lwd=3, xlab=mylab, bty="l")
    lines(densy, lty=2, col=2, lwd=3)
    xvals = c(densx$x, rev(densy$x))
    yvals = c(densx$y, rev(densy$y))
    polygon(xvals, yvals, col="gray")
  }
```

The `polygon()` function is used to fill in the area between the two curves. Results are shown in Figure 5.4).

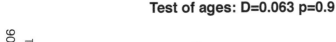

```
> mytitle = paste("Test of ages: D=", round(ksres$statistic, 3),
    " p=", round(ksres$p.value, 2), sep="")
> with(ds, plotdens(age[female==1], age[female==0], mytitle=mytitle,
    mylab="age (in years)"))
> legend(50, .05, legend=c("Women", "Men"), col=1:2, lty=1:2, lwd=2)
```

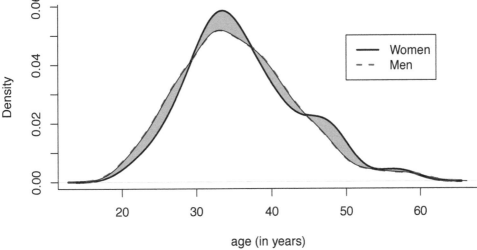

Figure 5.4: Density plot of age by gender

5.7.5 Survival analysis: logrank test

The logrank test (5.4.6) can be used to compare estimated survival curves between groups in the presence of censoring. Here we compare randomization groups with respect to `dayslink`, where a value of 0 for `linkstatus` indicates that the observation was censored, not observed, at the time recorded in `dayslink`.

```
> library(survival)
> survobj = survdiff(Surv(dayslink, linkstatus) ~ treat,
     data=ds)
> print(survobj)
Call:
survdiff(formula = Surv(dayslink, linkstatus) ~ treat, data = ds)

n=431, 22 observations deleted due to missingness.

          N Observed Expected (O-E)^2/E (O-E)^2/V
treat=0 209       35     92.8      36.0      84.8
treat=1 222      128     70.2      47.6      84.8

 Chisq= 84.8  on 1 degrees of freedom, p= 0
> names(survobj)
[1] "n"         "obs"        "exp"        "var"        "chisq"
[6] "na.action" "call"
```

Chapter 6

Linear regression and ANOVA

Regression and analysis of variance (ANOVA) form the basis of many investigations. Here we describe how to undertake many common tasks in linear regression (broadly defined), while Chapter 7 discusses many generalizations, including other types of outcome variables, longitudinal and clustered analysis, and survival methods.

Many R commands can perform linear regression, as it constitutes a special case of which many models are generalizations. We present detailed descriptions for the `lm()` command, as it offers the most flexibility and best output options tailored to linear regression in particular. While ANOVA can be viewed as a special case of linear regression, separate routines are available (`aov()`) to perform it.

R supports a flexible modeling language implemented using formulas (see `help(formula)` and 6.1.1) for regression that shares functionality with the lattice graphics functions (as well as other packages). Many of the routines available within R return or operate on `lm` class objects, which include objects such as coefficients, residuals, fitted values, weights, contrasts, model matrices, and similar quantities (see `help(lm)`).

The CRAN statistics for the social sciences task view provides an excellent overview of methods described here and in Chapter 7.

6.1   Model fitting

6.1.1   Linear regression

Example: 6.6.2

```
mod1 = lm(y ~ x1 + ... + xk, data=ds)
summary(mod1)
summary.aov(mod1)
```
or
```
form = as.formula(y ~ x1 + ... + xk)
mod1 = lm(form, data=ds)
summary(mod1)
coef(mod1)
```

Note: The first argument of the `lm()` function is a formula object, with the outcome specified followed by the ~ operator then the predictors. It returns a linear model object. More information about the linear model `summary()` command can be found using `help(summary.lm)`. The `coef()` function extracts coefficients from a model (see also the `coefplot` package). The `biglm()` function in the `biglm` package can support model fitting to very large datasets (see 6.1.7). By default, stars are used to annotate the output of

67

the `summary()` functions regarding significance levels: these can be turned off using the command `options(show.signif.stars=FALSE)`.

6.1.2   Linear regression with categorical covariates

Example: 6.6.2

See 6.1.4 (parameterization of categorical covariates).

```
ds = transform(ds, x1f = as.factor(x1))
mod1 = lm(y ~ x1f + x2 + ... + xk, data=ds)
```

Note: The `as.factor()` command creates a categorical variable from a variable. By default, the lowest value (either numerically or lexicographically) is the reference value. The `levels` option for the `factor()` function can be used to select a particular reference value (see 2.2.19). Ordered factors can be constructed using the `ordered()` function.

6.1.3   Changing the reference category

```
library(dplyr)
ds = mutate(ds, neworder = factor(classvar,
   levels=c("level", "otherlev1", "otherlev2")))
mod1 = lm(y ~ neworder, data=ds)
```

Note: The first level of a factor (by default, that which appears first lexicographically) is the reference group. This can be modified through use of the `factor()` function.

6.1.4   Parameterization of categorical covariates

Example: 6.6.6

The `as.factor()` function can be applied within any model-fitting command. Parameterization of the covariate can be controlled as below.

```
ds = transform(ds, x1f = as.factor(x1))
mod1 = lm(y ~ x1f, contrasts=list(x1f="contr.SAS"), data=ds)
```

Note: The `as.factor()` function creates a factor object. The `contrasts` option for the `lm()` function specifies how the levels of that factor object should be used within the function. The `levels` option to the `factor()` function allows specification of the ordering of levels (the default is lexicographic). An example can be found in Section 6.6.

The specification of the design matrix for analysis of variance and regression models can be controlled using the `contrasts` option. Examples of options (for a factor with four equally spaced levels) are given below.

```
> contr.treatment(4)              > contr.poly(4)
  2 3 4                                  .L   .Q     .C
1 0 0 0                           [1,] -0.671  0.5 -0.224
2 1 0 0                           [2,] -0.224 -0.5  0.671
3 0 1 0                           [3,]  0.224 -0.5 -0.671
4 0 0 1                           [4,]  0.671  0.5  0.224
> contr.SAS(4)                    > contr.sum(4)
  1 2 3                                [,1] [,2] [,3]
1 1 0 0                           1     1    0    0
2 0 1 0                           2     0    1    0
```

```
3 0 0 1                              3    0    0    1
4 0 0 0                              4   -1   -1   -1
> contr.helmert(4)
   [,1] [,2] [,3]
1   -1   -1   -1
2    1   -1   -1
3    0    2   -1
4    0    0    3
```

See `options("contrasts")` for defaults, and `contrasts()` or `C()` to apply a contrast function to a factor variable. Support for reordering factors is available within the `factor()` function.

6.1.5   Linear regression with no intercept

```
mod1 = lm(y ~ 0 + x1 + ... + xk, data=ds)
```
or
```
mod1 = lm(y ~ x1 + ... + xk - 1, data=ds)
```

6.1.6   Linear regression with interactions

Example: 6.6.2

```
mod1 = lm(y ~ x1 + x2 + x1:x2 + x3 + ... + xk, data=ds)
```
or
```
lm(y ~ x1*x2 + x3 + ... + xk, data=ds)
```

Note: The * operator includes all lower-order terms (in this case main effects), while the : operator includes only the specified interaction. So, for example, the commands y ~ x1*x2*x3 and y ~ x1 + x2 + x3 + x1:x2 + x1:x3 + x2:x3 + x1:x2:x3 are equivalent. The syntax also works with any covariates designated as categorical using the `as.factor()` command (see 6.1.2).

6.1.7   Linear regression with big data

```
library(biglm)
myformula = as.formula(y ~ x1)
res = biglm(myformula, chunk1)
res = update(res, chunk2)
coef(res)
```

Note: The `biglm()` and `update()` functions in the `biglm` package can fit linear (or generalized linear) models with dataframes larger than memory. It allows a single large model to be estimated in more manageable chunks, with results updated iteratively as each chunk is processed. The chunk size will depend on the application. The data argument may be a function, dataframe, `SQLiteConnection`, or `RODBC` connection object.

6.1.8   One-way analysis of variance

Example: 6.6.6

```
ds = transform(ds, xf=as.factor(x))
mod1 = aov(y ~ xf, data=ds)
summary(mod1)
anova(mod1)
```

Note: The `summary()` command can be used to provide details of the model fit. More information can be found using `help(summary.aov)`. Note that `summary.lm(mod1)` will display the regression parameters underlying the ANOVA model.

6.1.9   Analysis of variance with two or more factors

Example: 6.6.6

Interactions can be specified using the syntax introduced in 6.1.6 (see interaction plots, 8.5.2).

```
aov(y ~ as.factor(x1) + as.factor(x2), data=ds)
```

6.2   Tests, contrasts, and linear functions of parameters

6.2.1   Joint null hypotheses: several parameters equal 0

As an example, consider testing the null hypothesis $H_0 : \beta_1 = \beta_2 = 0$.

```
mod1 = lm(y ~ x1 + ... + xk, data=ds)
mod2 = lm(y ~ x3 + ... + xk, data=ds)
anova(mod2, mod1)
```

6.2.2   Joint null hypotheses: sum of parameters

As an example, consider testing the null hypothesis $H_0 : \beta_1 + \beta_2 = 1$.

```
mod1 = lm(y ~ x1 + ... + xk, data=ds)
covb = vcov(mod1)
coeff.mod1 = coef(mod1)
t = (coeff.mod1[2] + coeff.mod1[3] - 1)/
   sqrt(covb[2,2] + covb[3,3] + 2*covb[2,3])
pvalue = 2*(1-pt(abs(t), df=mod1$df))
```

6.2.3   Tests of equality of parameters

Example: 6.6.8

As an example, consider testing the null hypothesis $H_0 : \beta_1 = \beta_2$.

```
mod1 = lm(y ~ x1 + ... + xk, data=ds)
mod2 = lm(y ~ I(x1+x2) + ... + xk, data=ds)
anova(mod2, mod1)
```

or

```
library(gmodels)
estimable(mod1, c(0, 1, -1, 0, ..., 0))
```

or

```
mod1 = lm(y ~ x1 + ... + xk, data=ds)
covb = vcov(mod1)
coeff.mod1 = coef(mod1)
t = (coeff.mod1[2]-coeff.mod1[3])/sqrt(covb[2,2]+covb[3,3]-2*covb[2,3])
pvalue = 2*(1-pt(abs(t), mod1$df))
```

Note: The I() function inhibits the interpretation of operators, to allow them to be used as arithmetic operators. The estimable() function calculates a linear combination of the parameters. The more general code below utilizes the same approach introduced in 6.2.1 for the specific test of $\beta_1 = \beta_2$ (different coding would be needed for other comparisons).

6.2.4 Multiple comparisons

Example: 6.6.7

```
mod1 = aov(y ~ x, data=ds)
TukeyHSD(mod1, "x")
```

Note: The TukeyHSD() function takes an aov object as an argument and evaluates pairwise comparisons between all of the combinations of the factor levels of the variable x. (See the p.adjust() function, as well as the multcomp and factorplot packages for other multiple comparison methods, including Bonferroni, Holm, Hochberg, and false discovery rate adjustments.)

6.2.5 Linear combinations of parameters

Example: 6.6.8

It is often useful to find predicted values for particular covariate values. Here, we calculate the predicted value $E[Y|X_1 = 1, X_2 = 3] = \hat{\beta}_0 + \hat{\beta}_1 + 3\hat{\beta}_2$.

```
mod1 = lm(y ~ x1 + x2, data=ds)
newdf = data.frame(x1=c(1), x2=c(3))
predict(mod1, newdf, se.fit=TRUE, interval="confidence")
```
or
```
library(gmodels)
estimable(mod1, c(1, 1, 3))
```
or
```
library(mosaic)
myfun = makeFun(mod1)
myfun(x1=1, x2=3)
```

Note: The predict() command in R can generate estimates at any combination of parameter values, as specified as a dataframe that is passed as an argument. More information on this function can be found using help(predict.lm).

6.3 Model results and diagnostics

There are many functions available to produce predicted values and diagnostics. For additional commands not listed here, see help(influence.measures) and the "See also" in help(lm).

6.3.1 Predicted values

Example: 6.6.2

```
mod1 = lm(y ~ x, data=ds)
predicted.varname = predict(mod1)
```

Note: The command `predict()` operates on any `lm` object and by default generates a vector of predicted values.

6.3.2 Residuals

Example: 6.6.2

```
mod1 = lm(y ~ x, data=ds)
residual.varname = residuals(mod1)
```

Note: The command `residuals()` operates on any `lm` object and generates a vector of residuals. Other functions exist for `aov`, `glm`, or `lme` objects (see `help(residuals.glm)`).

6.3.3 Standardized and Studentized residuals

Example: 6.6.2

Standardized residuals are calculated by dividing the ordinary residual (observed minus expected, $y_i - \hat{y}_i$) by an estimate of its standard deviation. Studentized residuals are calculated in a similar manner, where the predicted value and the variance of the residual are estimated from the model fit while excluding that observation.

```
mod1 = lm(y ~ x, data=ds)
standardized.resid.varname = rstandard(mod1)
studentized.resid.varname = rstudent(mod1)
```

Note: The `rstandard()` and `rstudent()` functions operate on any `lm` object, and generate a vector of studentized residuals (the former command includes the observation in the calculation, while the latter does not).

6.3.4 Leverage

Example: 6.6.2

Leverage is defined as the diagonal element of the $(X(X^T X)^{-1} X^T)$ or "hat" matrix.

```
mod1 = lm(y ~ x, data=ds)
leverage.varname = hatvalues(mod1)
```

Note: The command `hatvalues()` operates on any `lm` object and generates a vector of leverage values.

6.3.5 Cook's distance

Example: 6.6.2

Cook's distance (D) is a function of the leverage (see 6.3.4) and the magnitude of the residual. It is used as a measure of the influence of a data point in a regression model.

```
mod1 = lm(y ~ x, data=ds)
cookd.varname = cooks.distance(mod1)
```

Note: The command `cooks.distance()` operates on any `lm` object and generates a vector of Cook's distance values.

6.3.6 DFFITs

Example: 6.6.2

DFFITs are a standardized function of the difference between the predicted value for the observation when it is included in the dataset and when (only) it is excluded from the dataset. They are used as an indicator of the observation's influence.

```
mod1 = lm(y ~ x, data=ds)
dffits.varname = dffits(mod1)
```

Note: The command `dffits()` operates on any `lm` object and generates a vector of DFFITS values.

6.3.7 Diagnostic plots

Example: 6.6.4

```
mod1 = lm(y ~ x, data=ds)
par(mfrow=c(2, 2)) # display 2 x 2 matrix of graphs
plot(mod1)
```

Note: The `plot.lm()` function (which is invoked when `plot()` is given a linear regression model as an argument) can generate six plots: (1) a plot of residuals against fitted values, (2) a Scale-Location plot of $\sqrt{(Y_i - \hat{Y}_i)}$ against fitted values, (3) a normal Q-Q plot of the residuals, (4) a plot of Cook's distances (6.3.5) versus row labels, (5) a plot of residuals against leverages (6.3.4), and (6) a plot of Cook's distances against leverage/(1−leverage). The default is to plot the first three and the fifth. The `which` option can be used to specify a different set (see `help(plot.lm)`).

6.3.8 Heteroscedasticity tests

```
library(lmtest)
bptest(y ~ x1 + ... + xk, data=ds)
```

Note: The `bptest()` function in the `lmtest` package performs the Breusch–Pagan test for heteroscedasticity [18]. Other diagnostic tests are available within the package.

6.4 Model parameters and results

6.4.1 Parameter estimates

Example: 6.6.2

```
mod1 = lm(y ~ x, data=ds)
coeff.mod1 = coef(mod1)
```

Note: The first element of the vector `coeff.mod1` is the intercept (assuming that a model with an intercept was fit).

6.4.2 Standardized regression coefficients

Standardized coefficients from a linear regression model are the parameter estimates obtained when the predictors and outcomes have been standardized to have a variance of 1 prior to model fitting.

```
library(QuantPsyc)
mod1 = lm(y ~ x)
lm.beta(mod1)
```

6.4.3   Coefficient plot

Example: 6.6.3

An alternative way to display regression results (coefficients and associated confidence intervals) is with a figure rather than a table [51].

```
library(mosaic)
mplot(mod, which=7)
```

Note: The specific coefficients to be displayed can be specified (or excluded, using negative values) via the `rows` option.

6.4.4   Standard errors of parameter estimates

See 6.4.10 (covariance matrix).

```
mod1 = lm(y ~ x, data=ds)
sqrt(diag(vcov(mod1)))
```
or
```
coef(summary(mod1))[,2]
```
Note: The standard errors are the second column of the results from `coef()`.

6.4.5   Confidence interval for parameter estimates

Example: 6.6.2

```
mod1 = lm(y ~ x, data=ds)
confint(mod1)
```

6.4.6   Confidence limits for the mean

These are the lower (and upper) confidence limits for the mean of observations with the given covariate values, as opposed to the prediction limits for individual observations with those values (see prediction limits, 6.4.7).

```
mod1 = lm(y ~ x, data=ds)
pred = predict(mod1, interval="confidence")
lcl.varname = pred[,2]
```

Note: The lower confidence limits are the second column of the results from `predict()`. To generate the upper confidence limits, the user would access the third column of the `predict()` object. The command `predict()` operates on any `lm()` object, and with these options generates confidence limit values. By default, the function uses the estimation dataset, but a separate dataset of values to be used to predict can be specified. The `panel=panel.lmbands` option from the `mosaic` package can be added to an `xyplot()` call to augment the scatterplot with confidence interval and prediction bands.

6.4.7   Prediction limits

These are the lower (and upper) prediction limits for "new" observations with the covariate values of subjects observed in the dataset, as opposed to confidence limits for the population mean (see confidence limits, 6.4.6).

```
mod1 = lm(y ~ ..., data=ds)
pred.w.lowlim = predict(mod1, interval="prediction")[,2]
```

Note: This code saves the second column of the results from the `predict()` function into a vector. To generate the upper confidence limits, the user would access the third column of the `predict()` object in R. The command `predict()` operates on any `lm()` object, and with these options generates prediction limit values. By default, the function uses the estimation dataset, but a separate dataset of values to be used to predict can be specified.

6.4.8   R-squared

```
mod1 = lm(y ~ ..., data=ds)
summary(mod1)$r.squared
```
or
```
library(mosaic)
rsquared(mod1)
```

6.4.9   Design and information matrix

See 3.3 (matrices).

```
mod1 = lm(y ~ x1 + ... + xk, data=ds)
XpX = t(model.matrix(mod1)) %*% model.matrix(mod1)
```
or
```
X = cbind(rep(1, length(x1)), x1, x2, ..., xk)
XpX = t(X) %*% X
rm(X)
```

Note: The `model.matrix()` function creates the design matrix from a linear model object. Alternatively, this quantity can be built up using the `cbind()` function to glue together the design matrix X. Finally, matrix multiplication (3.3.6) and the transpose function are used to create the information $(X'X)$ matrix.

6.4.10   Covariance matrix of parameter estimates

Example: 6.6.2

See 3.3 (matrices) and 6.4.4 (standard errors).

```
mod1 = lm(y ~ x, data=ds)
vcov(mod1)
```
or
```
sumvals = summary(mod1)
covb = sumvals$cov.unscaled*sumvals$sigma^2
```

Note: Running `help(summary.lm)` provides details on return values.

6.4.11   Correlation matrix of parameter estimates

See 3.3 (matrices) and 6.4.4 (standard errors).

```
mod1 = lm(y ~ x, data=ds)
mod1.cov = vcov(mod1)
mod1.cor = cov2cor(mod1.cov)
```

Note: The `cov2cor()` function is a convenient way to convert a covariance matrix into a correlation matrix.

6.5   Further resources

An accessible guide to linear regression in R can be found in [36]. Cook [28] reviews regression diagnostics. Frank Harrell's **rms** (regression modeling strategies) package [61] features extensive support for regression modeling. The CRAN statistics for the social sciences task view provides an excellent overview of methods described here and in Chapter 7.

6.6   Examples

To help illustrate the tools presented in this chapter, we apply many of the entries to the HELP data. The code can be downloaded from http://www.amherst.edu/~nhorton/r2/examples.

We begin by reading in the dataset and keeping only the female subjects. To prepare for future analyses, we create a version of substance as a factor variable (see 6.1.4) as well as dataframes containing subsets of our data.

```
> options(digits=3)
> # read in Stata format
> library(foreign)
> ds = read.dta("http://www.amherst.edu/~nhorton/r2/datasets/help.dta",
    convert.underscore=FALSE)
> library(dplyr)
> ds = mutate(ds, sub=factor(substance,
    levels=c("heroin", "alcohol", "cocaine")))
> newds = filter(ds, female==1)
> alcohol = filter(newds, substance=="alcohol")
> cocaine = filter(newds, substance=="cocaine")
> heroin = filter(newds, substance=="heroin")
```

6.6.1   Scatterplot with smooth fit

As a first step to help guide estimation of a linear regression, we create a scatterplot (8.3.1) displaying the relationship between age and the number of alcoholic drinks consumed in the period before entering detox (variable name: i1), as well as primary substance of abuse (alcohol, cocaine, or heroin).

Figure 6.1 displays a scatterplot of observed values for i1 (along with separate smooth fits by primary substance). To improve legibility, the plotting region is restricted to those with number of drinks between 0 and 40 (see plotting limits, 9.2.9).

```
> with(newds, plot(age, i1, ylim=c(0,40), type="n", cex.lab=1.2,
     cex.axis=1.2))
> with(alcohol, points(age, i1, pch="a"))
> with(alcohol, lines(lowess(age, i1), lty=1, lwd=2))
> with(cocaine, points(age, i1, pch="c"))
> with(cocaine, lines(lowess(age, i1), lty=2, lwd=2))
> with(heroin, points(age, i1, pch="h"))
> with(heroin, lines(lowess(age, i1), lty=3, lwd=2))
> legend(44, 38, legend=c("alcohol", "cocaine", "heroin"), lty=1:3,
     cex=1.4, lwd=2, pch=c("a", "c", "h"))
```

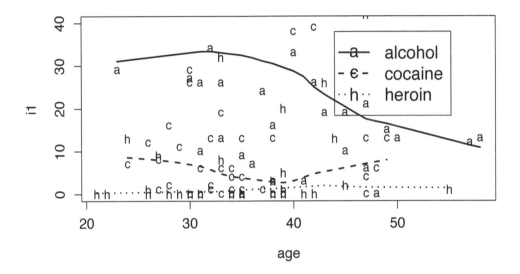

Figure 6.1: Scatterplot of observed values for age and I1 (plus smoothers by substance) using base graphics

The pch option to the legend() command can be used to insert plot symbols in R legends (Figure 6.1 displays the different line styles). A similar plot can be generated using the lattice package (see Figure 6.2). Finally, a third figure can be generated using the ggplot2 package (see Figure 6.3). Not surprisingly, the plots suggest a dramatic effect of primary substance, with alcohol users drinking more than others. There is some indication of an interaction with age.

6.6.2 Linear regression with interaction

Next we fit a linear regression model (6.1.1) for the number of drinks as a function of age, substance, and their interaction (6.1.6). We also fit the model with no interaction and use the anova() function to compare the models (the drop1() function could also be used).

```
> options(show.signif.stars=FALSE)
> lm1 = lm(i1 ~ sub * age, data=newds)
> lm2 = lm(i1 ~ sub + age, data=newds)
```

```
> xyplot(i1 ~ age, groups=substance, type=c("p", "smooth"),
    auto.key=list(columns=3, lines=TRUE, points=FALSE),
    ylim=c(0, 40), data=newds)
```

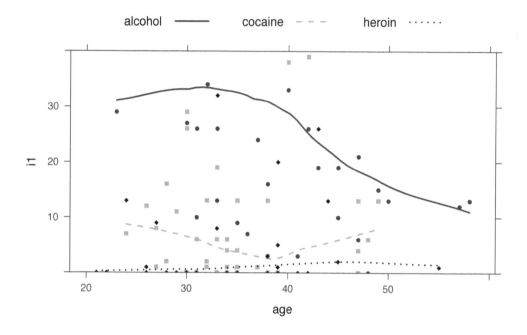

Figure 6.2: Scatterplot of observed values for age and I1 (plus smoothers by substance) using the lattice package

```
> anova(lm2, lm1)
Analysis of Variance Table

Model 1: i1 ~ sub + age
Model 2: i1 ~ sub * age
  Res.Df   RSS Df Sum of Sq    F Pr(>F)
1    103 26196
2    101 24815  2      1381 2.81  0.065
```

```
> summary.aov(lm1)
            Df Sum Sq Mean Sq F value  Pr(>F)
sub          2  10810    5405   22.00 1.2e-08
age          1     84      84    0.34   0.559
sub:age      2   1381     690    2.81   0.065
Residuals  101  24815     246
```

We observe a borderline significant interaction between age and substance group ($p = 0.065$). Additional information about the model can be displayed using the summary() and confint() functions.

```
> library(ggplot2)
> ggplot(data=newds, aes(x=age, y=i1)) + geom_point(aes(shape=substance)) +
    stat_smooth(method=loess, level=0.50, colour="black") +
    aes(linetype=substance) +
    coord_cartesian(ylim = c(0, 40)) +
    theme(legend.position="top") + labs(title="")
```

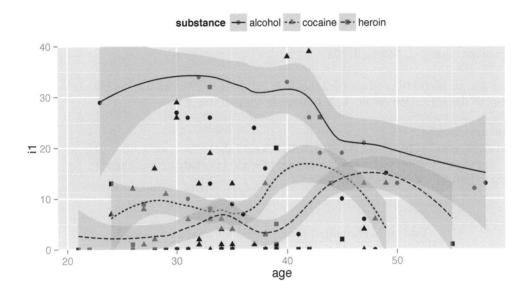

Figure 6.3: Scatterplot of observed values for age and I1 (plus smoothers by substance) using the `ggplot2` package

```
> summary(lm1)

Call:
lm(formula = i1 ~ sub * age, data = newds)

Residuals:
   Min    1Q Median    3Q    Max
-31.92  -8.25  -4.18   3.58  49.88

Coefficients:
              Estimate Std. Error t value Pr(>|t|)
(Intercept)     -7.770     12.879   -0.60  0.54763
subalcohol      64.880     18.487    3.51  0.00067
subcocaine      13.027     19.139    0.68  0.49763
age              0.393      0.362    1.09  0.28005
subalcohol:age  -1.113      0.491   -2.27  0.02561
subcocaine:age  -0.278      0.540   -0.51  0.60813
```

```
Residual standard error: 15.7 on 101 degrees of freedom
Multiple R-squared:  0.331,Adjusted R-squared:  0.298
F-statistic: 9.99 on 5 and 101 DF,  p-value: 8.67e-08
```

```
> confint(lm1)
                  2.5 %  97.5 %
(Intercept)     -33.319  17.778
subalcohol       28.207 101.554
subcocaine      -24.938  50.993
age              -0.325   1.112
subalcohol:age   -2.088  -0.138
subcocaine:age   -1.348   0.793
```

It may also be useful to produce the table in LaTeX format. We can use the xtable package to display the regression results in LaTeX as shown in Table 6.1.

```
> library(xtable)
> lmtab = xtable(lm1, digits=c(0,3,3,2,4), label="better",
>    caption="Formatted results using the {\\tt xtable} package")
> print(lmtab)  # output the LaTeX
```

Table 6.1: Formatted results using the xtable package

	Estimate	Std. Error	t value	Pr(>|t|)
(Intercept)	-7.770	12.879	-0.60	0.5476
subalcohol	64.880	18.487	3.51	0.0007
subcocaine	13.027	19.139	0.68	0.4976
age	0.393	0.362	1.09	0.2801
subalcohol:age	-1.113	0.491	-2.27	0.0256
subcocaine:age	-0.278	0.540	-0.51	0.6081

There are many quantities of interest stored in the linear model object lm1, and these can be viewed or extracted for further use.

```
> names(summary(lm1))
 [1] "call"           "terms"      "residuals"     "coefficients"
 [5] "aliased"        "sigma"      "df"            "r.squared"
 [9] "adj.r.squared"  "fstatistic" "cov.unscaled"
> summary(lm1)$sigma
[1] 15.7
```

```
> names(lm1)
 [1] "coefficients"   "residuals"  "effects"   "rank"
 [5] "fitted.values"  "assign"     "qr"        "df.residual"
 [9] "contrasts"      "xlevels"    "call"      "terms"
[13] "model"
```

```
> coef(lm1)
  (Intercept)       subalcohol       subcocaine            age subalcohol:age
       -7.770           64.880           13.027          0.393         -1.113
subcocaine:age
       -0.278
```

```
> vcov(lm1)
               (Intercept) subalcohol subcocaine     age subalcohol:age
(Intercept)         165.86    -165.86    -165.86  -4.548          4.548
subalcohol         -165.86     341.78     165.86   4.548         -8.866
subcocaine         -165.86     165.86     366.28   4.548         -4.548
age                  -4.55       4.55       4.55   0.131         -0.131
subalcohol:age        4.55      -8.87      -4.55  -0.131          0.241
subcocaine:age        4.55      -4.55     -10.13  -0.131          0.131
               subcocaine:age
(Intercept)             4.548
subalcohol             -4.548
subcocaine            -10.127
age                    -0.131
subalcohol:age          0.131
subcocaine:age          0.291
```

The entire table of regression coefficients and associated statistics can be saved as an object.

```
> mymodel = coef(summary(lm1))
> mymodel
               Estimate Std. Error t value Pr(>|t|)
(Intercept)      -7.770     12.879  -0.603 0.547629
subalcohol       64.880     18.487   3.509 0.000672
subcocaine       13.027     19.139   0.681 0.497627
age               0.393      0.362   1.086 0.280052
subalcohol:age   -1.113      0.491  -2.266 0.025611
subcocaine:age   -0.278      0.540  -0.514 0.608128
> mymodel[2,3]    # alcohol t-value
[1] 3.51
```

6.6.3   Regression coefficient plot

The mplot() function in the mosaic package generates a coefficient plot (6.4.3) for the main effects multiple regression model (see Figure 6.4).

6.6.4   Regression diagnostics

Assessing the model is an important part of any analysis. We begin by examining the residuals (6.3.2). First, we calculate the quantiles of their distribution (5.1.5), then display the smallest residual.

```
> library(mosaic)
> mplot(lm2, which=7, rows=-1)
[[1]]
```

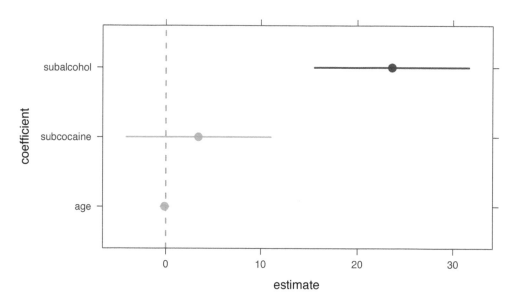

Figure 6.4: Regression coefficient plot

```
> library(dplyr)
> newds = mutate(newds, pred = fitted(lm1), resid = residuals(lm1))
> with(newds, quantile(resid))
    0%    25%    50%    75%   100%
-31.92  -8.25  -4.18   3.58  49.88
```

One way to print the largest value is to select the observation that matches the largest value. We use a series of "pipe" operations (A.5.3) to select a set of variables with the select() function, create the standardized residuals and add them to the dataset with the rstandard() function nested in the mutate() function, and then filter() out all rows except the one containing the maximum residual.

```
> library(dplyr)
> newds %>%
    select(id, age, i1, sub, pred, resid) %>%
    mutate(rstand = rstandard(lm1)) %>%
    filter(resid==max(resid))
  id age i1     sub pred resid rstand
1  9  50 71 alcohol 21.1  49.9   3.32
```

Graphical tools are one of the best ways to examine residuals. Figure 6.5 displays the default diagnostic plots (6.3) from the model.

Figure 6.6 displays the empirical density of the standardized residuals, along with an

```
> oldpar = par(mfrow=c(2, 2), mar=c(4, 4, 2, 2) + .1)
> plot(lm1); par(oldpar)
```

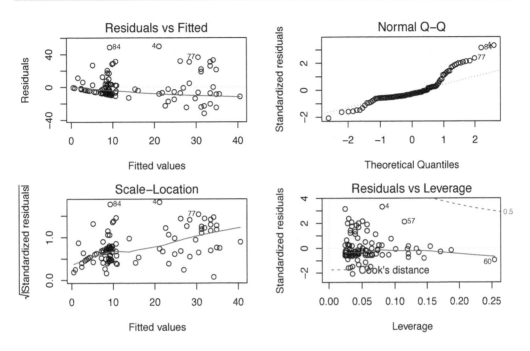

Figure 6.5: Default diagnostics for linear models

overlaid normal density. The assumption that the residuals are approximately Gaussian does not appear to be tenable.

The residual plots also indicate some potentially important departures from model assumptions: further exploration and model assessment should be undertaken.

6.6.5 Fitting a regression model separately for each value of another variable

One common task is to perform identical analyses in several groups. Here, as an example, we consider separate linear regressions for each substance abuse group.

A matrix of the correct size is created, then a `for` loop is run for each unique value of the grouping variable.

```
> uniquevals = unique(newds$substance)
> numunique = length(uniquevals)
> formula = as.formula(i1 ~ age)
> p = length(coef(lm(formula, data=newds)))
> res = matrix(rep(0, numunique*p), p, numunique)
> for (i in 1:length(uniquevals)) {
      res[,i] = coef(lm(formula,
          data=subset(newds, substance==uniquevals[i])))
  }
> rownames(res) = c("intercept","slope")
> colnames(res) = uniquevals
```

```
> library(MASS)
> std.res = rstandard(lm1)
> hist(std.res, breaks=seq(-2.5, 3.5, by=.5), main="",
    xlab="standardized residuals", col="gray80", freq=FALSE)
> lines(density(std.res), lwd=2)
> xvals = seq(from=min(std.res), to=max(std.res), length=100)
> lines(xvals, dnorm(xvals, mean(std.res), sd(std.res)), lty=2)
```

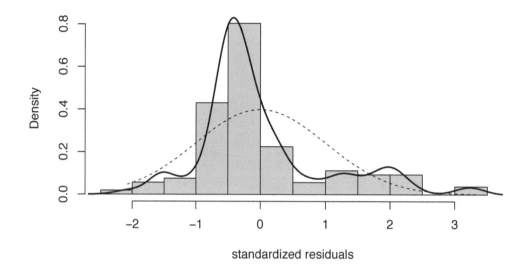

Figure 6.6: Empirical density of residuals, with superimposed normal density

```
> res
          heroin cocaine alcohol
intercept -7.770   5.257   57.11
slope      0.393   0.116   -0.72
```

6.6.6   Two-way ANOVA

Is there a statistically significant association between gender and substance abuse group with depressive symptoms? An interaction plot (8.5.2) may be helpful in making a determination. The `interaction.plot()` function can be used to carry out this task. Figure 6.7 displays an interaction plot for CESD as a function of substance group and gender.

```
> library(dplyr)
> ds = mutate(ds, genf = as.factor(ifelse(female, "F", "M")))
```

There are indications of large effects of gender and substance group, but little suggestion of interaction between the two. The same conclusion is reached in Figure 6.8, which displays boxplots by substance group and gender. We begin by creating better labels for the grouping variable, using the `cases()` function from the `memisc` package.

```
> with(ds, interaction.plot(substance, genf, cesd,
    xlab="substance", las=1, lwd=2))
```

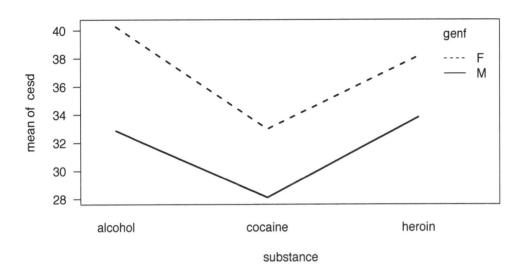

Figure 6.7: Interaction plot of CESD as a function of substance group and gender

```
> library(dplyr)
> library(memisc)
> ds = mutate(ds, subs = cases(
    "Alc" = substance=="alcohol",
    "Coc" = substance=="cocaine",
    "Her" = substance=="heroin"))
```

The width of each box is proportional to the size of the sample, with the notches denoting confidence intervals for the medians and X's marking the observed means. Next, we proceed to formally test whether there is a significant interaction through a two-way analysis of variance (6.1.9). We fit models with and without an interaction, and then compare the results. We also construct the likelihood ratio test manually.

```
> aov1 = aov(cesd ~ sub * genf, data=ds)
> aov2 = aov(cesd ~ sub + genf, data=ds)
> anova(aov2, aov1)
Analysis of Variance Table

Model 1: cesd ~ sub + genf
Model 2: cesd ~ sub * genf
  Res.Df   RSS Df Sum of Sq   F Pr(>F)
1    449 65515
2    447 65369  2       146 0.5   0.61
```

```
> boxout = with(ds,
      boxplot(cesd ~ subs + genf, notch=TRUE, varwidth=TRUE,
        col="gray80"))
> boxmeans = with(ds, tapply(cesd, list(subs, genf), mean))
> points(seq(boxout$n), boxmeans, pch=4, cex=2)
```

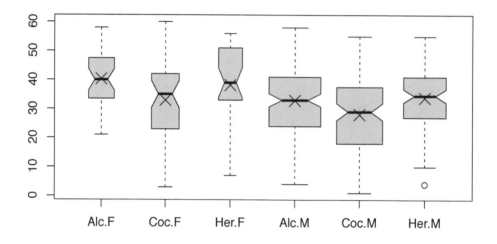

Figure 6.8: Boxplot of CESD as a function of substance group and gender

```
> options(digits=8)
> logLik(aov1)
'log Lik.' -1768.9186 (df=7)
> logLik(aov2)
'log Lik.' -1769.4236 (df=5)
> lldiff = logLik(aov1)[1] - logLik(aov2)[1]
> lldiff
[1] 0.50505522
> 1 - pchisq(2*lldiff, df=2)
[1] 0.60347225
> options(digits=3)
```

There is little evidence ($p > 0.6$) of an interaction, so this term can be dropped.

```
> summary(aov2)
            Df Sum Sq Mean Sq F value  Pr(>F)
sub          2   2704    1352    9.27 0.00011
genf         1   2569    2569   17.61 3.3e-05
Residuals  449  65515     146
```

The AIC (Akaike Information Criterion) statistic (7.8.3) can also be used to compare models.

```
> AIC(aov1)
[1] 3552
> AIC(aov2)
[1] 3549
```

The AIC criterion also suggests that the model without the interaction is most appropriate.

It may be useful to change the default reference level for variables. The default R design matrix (see 6.1.4) can be changed and the model re-fit.

```
> contrasts(ds$sub) = contr.SAS(3)
> aov3 = lm(cesd ~ sub + genf, data=ds)
> summary(aov3)

Call:
lm(formula = cesd ~ sub + genf, data = ds)

Residuals:
   Min    1Q Median    3Q    Max
-32.13 -8.85   1.09  8.48  27.09

Coefficients:
            Estimate Std. Error t value Pr(>|t|)
(Intercept)    33.52       1.38   24.22  < 2e-16
sub1            5.61       1.46    3.83  0.00014
sub2            5.32       1.34    3.98  8.1e-05
genfM          -5.62       1.34   -4.20  3.3e-05

Residual standard error: 12.1 on 449 degrees of freedom
Multiple R-squared:  0.0745,Adjusted R-squared:  0.0683
F-statistic:    12 on 3 and 449 DF,  p-value: 1.35e-07
```

6.6.7 Multiple comparisons

We can also carry out multiple comparison (6.2.4) procedures to test each of the pairwise differences between substance abuse groups, using the `TukeyHSD()` function.

```
> mult = TukeyHSD(aov(cesd ~ sub, data=ds), "sub")
> mult
  Tukey multiple comparisons of means
    95% family-wise confidence level

Fit: aov(formula = cesd ~ sub, data = ds)

$sub
                  diff   lwr   upr p adj
alcohol-heroin  -0.498 -3.89  2.89 0.936
cocaine-heroin  -5.450 -8.95 -1.95 0.001
cocaine-alcohol -4.952 -8.15 -1.75 0.001
```

The alcohol group and heroin group both have significantly higher CESD scores than the cocaine group, but the alcohol and heroin groups do not significantly differ from each other (95% confidence interval (CI) for the difference ranges from -3.9 to 2.9). Figure 6.9 provides a graphical display of the pairwise comparisons.

The `factorplot()` function in the `factorplot` package provides an alternative plotting scheme. This is demonstrated using a model where the CESD scores are grouped into six categories.

```
> require(mosaic)
> mplot(mult)
```

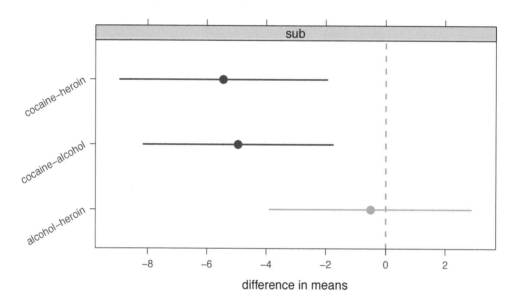

Figure 6.9: Pairwise comparisons (using Tukey HSD procedure)

```
> library(dplyr)
> library(factorplot)
> newds = mutate(newds, cesdgrp = cut(cesd,
    breaks=c(-1, 10, 20, 30, 40, 50, 61),
    labels=c("0-10", "11-20", "21-30", "31-40", "41-50", "51-60")))
> tally(~ cesdgrp, data=newds)

 0-10 11-20 21-30 31-40 41-50 51-60
    4    10    18    31    24    20
> mod = lm(pcs ~ age + cesdgrp, data=newds)
> fp = factorplot(mod, adjust.method="none", factor.variable="cesdgrp",
    pval=0.05, two.sided=TRUE, order="natural")
```

Figure 6.10 provides a graphical display of the fifteen pairwise comparisons, where the pairwise difference is displayed above the standard error of that difference (in italics).

6.6.8   Contrasts

We can also fit contrasts (6.2.3) to test hypotheses involving multiple parameters. In this case, we can compare the CESD scores for the alcohol and heroin groups to the cocaine group.

```
> plot(fp, abbrev.char=100)
```

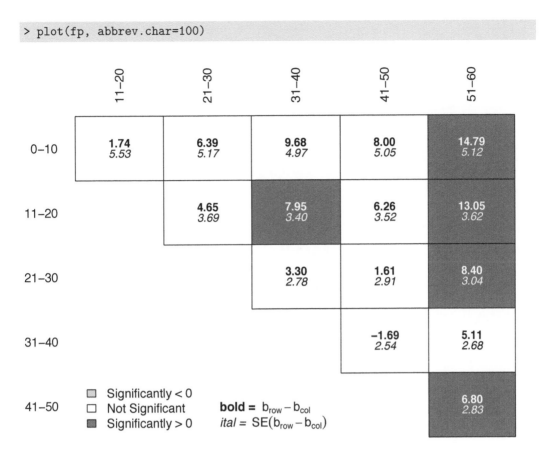

Figure 6.10: Pairwise comparisons (using the factorplot function)

```
> library(gmodels)
> levels(ds$sub)
[1] "heroin"  "alcohol" "cocaine"
> fit.contrast(aov2, "sub", c(1,1,-2), conf.int=0.95 )
               Estimate Std. Error t value Pr(>|t|) lower CI upper CI
sub c=( 1 1 -2 )     10.9        2.42    4.52 8.04e-06     6.17     15.7
```

As expected from the interaction plot (Figure 6.7), there is a statistically significant difference in this 1-degree-of-freedom comparison ($p < 0.0001$).

Chapter 7

Regression generalizations and modeling

This chapter extends the discussion of linear regression introduced in Chapter 6 to include many commonly used statistical methods and models. The CRAN statistics for the social sciences task view provides an excellent overview of methods described here and in Chapter 6.

7.1 Generalized linear models

Table 7.1 displays the options to specify link functions and family of distributions for generalized linear models [111]. Description of several specific generalized linear regression models (e.g., logistic and Poisson) can be found in subsequent sections of this chapter.

```
glmod1 = glm(y ~ x1 + ... + xk, family="familyname"(link="linkname"),
    data=ds)
```

Note: More information on GLM families and links can be found using `help(family)`. Nested models can be compared using `anova(mymod2, mymod1, test="Chisq")`.

7.1.1 Logistic regression model

Example: 7.10.1

```
glm(y ~ x1 + ... + xk, binomial, data=ds)
```
or
```
library(rms)
lrm(y ~ x1 + ... + xk, data=ds)
```

Note: The `lrm()` function within the `rms` package provides the so-called "c" statistic (area under receiver operating characteristic curve, see 8.5.7) and the Nagelkerke pseudo-R^2 index [120]. Nested models can be compared using `anova(mymod2, mymod1, test="Chisq")`.

7.1.2 Conditional logistic regression model

```
library(survival)
cmod = clogit(y ~ x1 + ... + xk + strata(id), data=ds)
```

Table 7.1: Generalized linear model distributions supported

Distribution	R glm()
Gaussian	family="gaussian", link="identity", "log" or "inverse"
binomial	family="binomial", link="logit", "probit", "cauchit", "log" or "cloglog"
gamma	family="Gamma", link="inverse", "identity" or "log"
Poisson	family="poisson", link="log", "identity" or "sqrt"
inverse Gaussian	family="inverse.gaussian", link="1/mu^2", "inverse", "identity" or "sqrt"
multinomial	See multinom() in nnet package
negative binomial	See negative.binomial() in MASS package
overdispersed	family="quasi", link="logit", "probit", "cloglog", "identity", "inverse", "log", "1/mu^2" or "sqrt" (see glm.binomial.disp() in the dispmod package)

Note: For the glm() function, the available links for each distribution are listed.

Note: The variable id identifies strata or matched sets of observations. An exact model is fit by default.

7.1.3   Exact logistic regression

```
library(elrm)
ds = transform(ds, n=1)
elrmres = elrm(y/n ~ x1 + ... + xk, interest=~x1, iter=1100,
  burnIn=100, data=ds)
```

Note: The elrm() function implements a modified MCMC (Markov Chain Monte Carlo) algorithm to approximate exact conditional inference for logistic regression models [202]. The binomial response must be provided in the form y/n, where y specifies the number of successes and n indicates the number of binomial trials for each row of the dataframe.

7.1.4   Ordered logistic model

Example: 7.10.6

In this model, the odds of each level of the outcome relative to all lower levels are calculated. A key assumption of the model is that the odds are proportional across levels.

```
library(MASS)
polr(y ~ x1 + ... + xk, data=ds)
```

Note: The default link is logistic; this can be changed to probit, complementary log-log, or Cauchy using the method option (see also ordered()).

7.1.5 Generalized logistic model

Example: 7.10.7

```
library(VGAM)
mlogit = vglm(y ~ x1 + ... + xk, family=multinomial(), data=ds)
```

7.1.6 Poisson model

Example: 7.10.2

See 7.2.1 (zero-inflated Poisson).

```
glm(y ~ x1 + ... + xk, poisson, data=ds)
```

Note: It is always important to check assumptions for models. This is particularly true for Poisson models, which are quite sensitive to model departures [69]. One way to assess the fit of the model is by comparing the observed and expected cell counts, and then calculating Pearson's chi-square statistic. This can be carried out using the `goodfit()` function.

7.1.7 Negative binomial model

Example: 7.10.4

See 7.2.2 (zero-inflated negative binomial).

```
library(MASS)
glm.nb(y ~ x1 + ... + xk, data=ds)
```

7.1.8 Log-linear model

Log-linear models are a flexible approach to analysis of categorical data [3]. A log-linear model of a three-dimensional contingency table denoted by X_1, X_2, and X_3 might assert that the expected counts depend on a two-way interaction between the first two variables, but that X_3 is independent of all the others:

$$log(m_{ijk}) = \mu + \lambda_i^{X_1} + \lambda_j^{X_2} + \lambda_{ij}^{X_1,X_2} + \lambda_k^{X_3}$$

```
logres = loglin(table(x1, x2, x3), margin=list(c(1,2), c(3)), param=TRUE)
pvalue = 1 - with(logres, pchisq(lrt, df))
```

Note: The `margin` option specifies the dependence assumptions. In addition to the `loglin()` function, the `loglm()` function within the `MASS` package provides an interface for log-linear modeling.

7.2 Further generalizations

7.2.1 Zero-inflated Poisson model

Example: 7.10.3

Zero-inflated Poisson models can be used for count outcomes that generally follow a Poisson distribution but for which there are (many) more observed counts of 0 than would be expected. These data can be seen as deriving from a mixture distribution of a Poisson and a degenerate distribution with point mass at zero (see 7.2.2, zero-inflated negative binomial).

```
library(pscl)
mod = zeroinfl(y ~ x1 + ... + xk | x2 + ... + xp, data=ds)
```

Note: The Poisson rate parameter of the model is specified in the usual way with a formula as argument to `zeroinfl()`. The default link is `log`. The zero probability is modeled as a function of the covariates specified after the "|" character. An intercept-only model can be fit by including 1 as the second model. Support for zero-inflated negative binomial and geometric models is available.

7.2.2   Zero-inflated negative binomial model

Zero-inflated negative binomial models can be used for count outcomes that generally follow a negative binomial distribution but for which there are (many) more observed counts of 0 than would be expected. These data can be seen as deriving from a mixture distribution of a negative binomial and a degenerate distribution with point mass at zero (see zero-inflated Poisson, 7.2.1).

```
library(pscl)
mod = zeroinfl(y ~ x1 + ... + xk | x2 + ... + xp, dist="negbin", data=ds)
```

Note: The negative binomial rate parameter of the model is specified in the usual way with a formula as argument to `zeroinfl()`. The default link is `log`. The zero probability is modeled as a function of the covariates specified after the '|' character. A single intercept for all observations can be fit by including 1 as the model.

7.2.3   Generalized additive model

Example: 7.10.8

```
library(gam)
gam(y ~ s(x1, df) + lo(x2) + lo(x3, x4) + x5 + ... + xk, data=ds)
```

Note: Specification of a smooth term for variable `x1` is given by `s(x1)`, while a univariate or bivariate loess fit can be included using `lo(x1)` or `lo(x1, x2)`. See `gam.s()` and `gam.lo()` within the `gam` package for details regarding specification of degrees of freedom or span, respectively. Polynomial regression terms can be fit using the `poly()` function.

7.2.4   Nonlinear least squares model

Nonlinear least squares models [156] can be fit. As an example, consider the income inequality model described by Sarabia and colleagues [146]:

$$Y = (1 - (1 - X)^p)^{(1/p)}$$

```
nls(y ~ (1- (1-x)^{p})^(1/{p}), start=list(p=0.5), trace=TRUE)
```

Note: We provide a starting value (0.5) within the interior of the parameter space. Finding solutions for nonlinear least squares problems is often challenging (consult `help(nls)` for information on supported algorithms as well as Section 3.2.9, optimization).

7.3 Robust methods

7.3.1 Quantile regression model

Example: 7.10.5

Quantile regression predicts changes in the specified quantile of the outcome variable per unit change in the predictor variables, analogous to the change in the mean predicted in least squares regression. If the quantile so predicted is the median, this is equivalent to minimum absolute deviation regression (as compared to least squares regression minimizing the squared deviations).

```
library(quantreg)
quantmod = rq(y ~ x1 + ... + xk, tau=0.75, data=ds)
```

Note: The default for `tau` is 0.5, corresponding to median regression. If a vector is specified, the return value includes a matrix of results.

7.3.2 Robust regression model

Robust regression refers to methods for detecting outliers and/or providing stable estimates when they are present. Outlying variables in the outcome, predictor, or both are considered.

```
library(MASS)
rlm(y ~ x1 + ... + xk, data=ds)
```

Note: The `rlm()` function fits a robust linear model using M estimation. More information can be found in the CRAN robust statistical methods task view.

7.3.3 Ridge regression model

Ridge regression is used to deal with ill-conditional regression problems, particularly those due to multicollinearity (see also 7.8.5).

```
library(MASS)
ridgemod = lm.ridge(y ~ x1 + ... + xk, lambda=seq(from=a, to=b, by=c),
    data=ds)
```

Note: Post-estimation functions supporting `ridgelm` objects include `plot()` and `select()`. A vector of ridge constants can be specified using the `lambda` option.

7.4 Models for correlated data

There is extensive support for correlated data regression models, including repeated measures, longitudinal, time series, clustered, and other related methods. Throughout this section, we assume that repeated measurements are taken on a subject or cluster with a common value for the variable `id`.

7.4.1   Linear models with correlated outcomes

Example: 7.10.10

```
library(nlme)
glsres = gls(y ~ x1 + ... + xk,
   correlation=corSymm(form = ~ ordervar | id),
   weights=varIdent(form = ~1 | ordervar), ds)
```

Note: The `gls()` function supports estimation of generalized least squares regression models with arbitrary specification of the variance covariance matrix. In addition to a formula interface for the mean model, the analyst specifies a within-group correlation structure as well as a description of the within-group heteroscedasticity structure (using the `weights` option). The statement `ordervar | id` implies that associations are assumed within `id`. Other covariance matrix options are available; see `help(corClasses)`.

7.4.2   Linear mixed models with random intercepts

See 7.4.3 (random slope models), 7.4.4 (random coefficient models), and 11.2 (empirical power calculations).

```
library(nlme)
lmeint = lme(fixed= y ~ x1 + ... + xk, random = ~ 1 | id,
   na.action=na.omit, data=ds)
```

Note: Best linear unbiased predictors (BLUPs) of the sum of the fixed effects plus corresponding random effects can be generated using the `coef()` function, random effect estimates using the `random.effects()` function, and the estimated variance–covariance matrix of the random effects using `VarCorr()` (see `fixef()` and `ranef()`). Normalized residuals (using a Cholesky decomposition, see pages 238–241 of Fitzmaurice et al. [40]) can be generated using the `type="normalized"` option when calling `residuals()` using an NLME option (more information can be found using `help(residuals.lme)`). A plot of the random effects can be created using `plot(lmeint)`. See the `lmmfit` package for goodness-of-fit measures for linear mixed models with one level of clustering.

7.4.3   Linear mixed models with random slopes

Example: 7.10.11

See 7.4.2 (random intercept models) and 7.4.4 (random coefficient models).

```
library(nlme)
lmeslope = lme(fixed=y ~ time + x1 + ... + xk, random = ~ time | id,
   na.action=na.omit, data=ds)
```

Note: The default covariance for the random effects is unstructured (see `help(reStruct)` for other options). Best linear unbiased predictors (BLUPs) of the sum of the fixed effects plus corresponding random effects can be generated using the `coef()` function, random effect estimates using the `random.effects()` function, and the estimated variance covariance matrix of the random effects using `VarCorr()`. A plot of the random effects can be created using `plot(lmeint)`.

7.4.4 More complex random coefficient models

We can extend the random effects models introduced in 7.4.2 and 7.4.3 to three or more subject-specific random parameters (e.g., a quadratic growth curve or spline/"broken stick" model [40]). We use $time_1$ and $time_2$ to refer to two generic functions of time.

```
library(nlme)
lmestick = lme(fixed= y ~ time1 + time2 + x1 + ... + xk,
   random = ~ time1 time2 | id, data=ds, na.action=na.omit)
```

Note: The default covariance for the random effects is unstructured (see `help(reStruct)` for other options). Best linear unbiased predictors (BLUPs) of the sum of the fixed effects plus corresponding random effects can be generated using the `coef()` function, random effect estimates using the `random.effects()` function, and the estimated variance covariance matrix of the random effects using `VarCorr()`. A plot of the random effects can be created using `plot(lmeint)`.

7.4.5 Multilevel models

Studies with multiple levels of clustering can be estimated. In a typical example, a study might include schools (as one level of clustering) and classes within schools (a second level of clustering), with individual students within the classrooms providing a response. Generically, we refer to $level_l$ variables, which are identifiers of cluster membership at level l. Random effects at different levels are assumed to be uncorrelated with each other.

```
library(nlme)
lmres = lme(fixed= y ~ x1 + ... + xk, random= ~ 1 | level1 / level2,
   data=ds)
```

Note: A model with k levels of clustering can be fit using the syntax: `level1 / ... / levelk`.

7.4.6 Generalized linear mixed models

Examples: 7.10.13 and 11.2

```
library(lme4)
glmmres = glmer(y ~ x1 + ... + xk + (1|id), family=familyval, data=ds)
```

Note: See `help(family)` for details regarding specification of distribution families and link functions.

7.4.7 Generalized estimating equations

Example: 7.10.12

```
library(gee)
geeres = gee(formula = y ~ x1 + ... + xk, id=id, data=ds,
   family=binomial, corstr="independence")
```

Note: The `gee()` function requires that the dataframe be sorted by subject identifier. Other correlation structures include `"fixed"`, `"stat_M_dep"`, `"non_stat_M_dep"`, `"AR-M"`, and `"unstructured"`. Note that the `"unstructured"` working correlation requires careful specification of ordering when missing data are monotone.

7.4.8 MANOVA

```
library(car)
mod = lm(cbind(y1, y2, y3) ~ x1, data=ds)
Anova(mod, type="III")
```

Note: The `car` package has a vignette that provides detailed examples, including a repeated measures ANOVA with details of use of the `idata` and `idesign` options. If the factor `x1` has two levels, this is the equivalent of a Hotelling's T^2 test. The `Hotelling` package (due to James Curran) can also be used to calculate Hotelling's $T2$ statistic.

7.4.9 Time series model

Time series modeling is an extensive area with a specialized language and notation. We make only the briefest approach here. We display fitting an ARIMA (autoregressive integrated moving average) model for the first difference, with first-order autoregression and moving averages. The CRAN time series task view provides an overview of support available for R.

```
tsobj = ts(x, frequency=12, start=c(1992, 2))
arres = arima(tsobj, order=c(1, 1, 1))
```

Note: The `ts()` function creates a time series object, in this case for monthly time series data within the variable `x` beginning in February 1992 (the default behavior is that the series starts at time 1 and the number of observations per unit of time is 1). The `start` option is either a single number or a vector of two integers that specify a natural time unit and a number of samples into the time unit. The `arima()` function fits an ARIMA model with AR, differencing, and MA order, all equal to 1.

7.5 Survival analysis

Survival, or failure time data, typically consist of the time until the event, as well as an indicator of whether the event was observed or censored at that time. Throughout, we denote the time of measurement with the variable `time` and censoring with a dichotomous variable `cens` $= 1$ if censored, or $= 0$ if observed. More information on survival (or failure time, or time-to-event) analysis can be found in the CRAN survival analysis task view (see A.6.4). Other entries related to survival analysis include 5.4.6 (log-rank test) and 8.5.11 (Kaplan–Meier plot).

7.5.1 Proportional hazards (Cox) regression model

Example: 7.10.14

```
library(survival)
survmod = coxph(Surv(time, cens) ~ x1 + ... + xk)
```

Note: The Efron estimator is the default; other choices including exact and Breslow can be specified using the `method` option. The `cph()` function within the `rms` package supports time-varying covariates, while the `cox.zph()` function within the `survival` package allows testing of the proportionality assumption, as does the `simPH` package.

7.5.2 Proportional hazards (Cox) model with frailty

```
library(survival)
coxph(Surv(time, cens) ~ x1 + ... + xk + frailty(id), data=ds)
```

Note: More information on specification of frailty models can be found using `help(frailty)`; support is available for t, Gamma, and Gaussian distributions.

7.5.3 Nelson–Aalen estimate of cumulative hazard

The Nelson–Aalen method provides a non-parametric estimator of the cumulative hazard rate function in censored data problems [186].

```
calcna = function(time, event) {
   na.fit = survfit(coxph(Surv(time, event) ~ 1), type="aalen")
   jumps = c(0, na.fit$time, max(time))
   # need to be careful at the beginning and end
   surv = c(1, na.fit$surv, na.fit$surv[length(na.fit$surv)])

   # apply appropriate transformation
   neglogsurv = -log(surv)

   # create placeholder of correct length
   naest = numeric(length(time))
   for (i in 2:length(jumps)) {
      naest[which(time>=jumps[i-1] & time<=jumps[i])] =
         neglogsurv[i-1]   # select the appropriate value
   }
   return(naest)
}
```
or
```
basehaz(coxph(Surv(time, event) ~ 1))
```

Note: We can do the necessary housekeeping, using the fact that the Nelson–Aalen estimate is just the negative log of the survival function (after specifying the `type="aalen"` option. Similar estimates can be generated using the `basehaz()` function.

7.5.4 Testing the proportionality of the Cox model

There are several methods for assessing whether the proportionality assumption holds.

```
library(survival)
survmod = coxph(Surv(time, cens) ~ x1 + ... + xk)
cox.zph(survmod)
plot(cox.zph(survmod))
```

Note: The `cox.zph()` function supports a plot object to generate graphical displays that facilitate model assessment (see also John Fox's Cox regression appendix at http://tinyurl.com/foxcox).

7.5.5   Cox model with time-varying predictors

Estimation of the Cox model with time-varying predictors requires the creation of a dataset with separate time periods for each occurrence of the time-varying predictor. More details can be found in John Fox's Cox regression appendix at `http://tinyurl.com/foxcox`.

7.6   Multivariate statistics and discriminant procedures

This section describes some commonly used multivariate, clustering methods, and discriminant procedures [109, 164]. The multivariate statistics, cluster analysis, and psychometrics task views on CRAN provide additional descriptions of available functionality.

7.6.1   Cronbach's α

Example: 7.10.15

Cronbach's α is a measure of internal consistency for a multi-item measure.

```
library(multilevel)
cronbach(cbind(x1, x2, ..., xk))
```

7.6.2   Factor analysis

Example: 7.10.16

Factor analysis is used to explain the variability of a set of measures in terms of underlying unobservable factors. The observed measures can be expressed as linear combinations of the factors plus random error. Factor analysis is often used as a way to guide the creation of summary scores from individual items.

```
res = factanal(~ x1 + ... + xk, factors=3, scores="regression")
print(res, cutoff=0.45, sort=TRUE)
```

7.6.3   Recursive partitioning

Example: 7.10.17

Recursive partitioning is used to create a decision tree to classify observations from a dataset based on categorical predictors.

```
library(rpart)
mod.rpart = rpart(y ~ x1 + ... + xk, method="class", data=ds)
printcp(mod.rpart)
plot(mod.rpart)
text(mod.rpart)
```

Note: The `partykit` package provides more control of the display of regression trees (see also the CRAN machine learning task view).

7.6.4   Linear discriminant analysis

Example: 7.10.18

Linear (or Fisher) discriminant analysis is used to find linear combinations of variables that can predict class membership.

```
library(MASS)
ngroups = length(levels(group))
ldamodel = lda(y ~ x1 + ... + xk, prior=rep(1/ngroups, ngroups))
print(ldamodel)
```

7.6.5   Latent class analysis

Latent class analysis is a technique used to classify observations based on patterns of categorical responses.

```
library(poLCA)
poLCA(cbind(x1, x2, ..., x3) ~ 1, maxiter=50000, nclass=k, nrep=n, data=ds)
```

Note: In this example, a k class model is fit. The `poLCA()` function requires that the variables are coded as positive integers. Other support for latent class models can be found in the `randomLCA` and the `MplusAutomation` packages.

7.6.6   Hierarchical clustering

Example: 7.10.19

Many techniques exist for grouping similar variables or similar observations. These groups, or clusters, can be overlapping or disjoint, and are sometimes placed in a hierarchical structure so that some disjoint clusters share a higher-level cluster. Clustering tools in the `stats` package include `hclust()` and `kmeans()`. The function `dendrogram()`, also in the `stats` package, plots tree diagrams. The `cluster()` package contains functions `pam()`, `clara()`, and `diana()`. The CRAN clustering task view has more details.

```
cormat = cor(cbind(x1, x2, ..., xk), use="pairwise.complete.obs")
hclustobj = hclust(dist(cormat))
```

7.7   Complex survey design

The appropriate analysis of sample surveys requires incorporation of complex design features, including stratification, clustering, weights, and finite population correction. These can be addressed for many common models. In this example, we assume that there are variables `psuvar` (cluster or primary sampling unit), `stratum` (stratification variable), and `wt` (sampling weight). Code examples are given to estimate the mean of a variable `x1` as well as a linear regression model.

```
library(survey)
mydesign = svydesign(id=~psuvar, strata=~stratum, weights=~wt,
   fpc=~fpcvar, data=ds)
meanres = svymean(~ x1, mydesign)
regres = svyglm(y ~ x1 + ... + xk, design=mydesign)
```

Note: Thomas Lumley's `survey` package includes support for many models. Illustrated above are means and linear regression models, with specification of PSUs, stratification, weight, and finite population correction. The CRAN official statistics task view provides an overview of other implementations.

7.8   Model selection and assessment

7.8.1   Compare two models

Example: 6.6.6

The function `drop1()` computes a table of changes in fit. In addition, nested modes can be compared using the `anova()` function.

```
mod1 = lm(y ~ x1 + ... + xk, data=ds)
mod2 = lm(y ~ x3 + ... + xk, data=ds)
anova(mod2, mod1)
```
or
```
drop1(mod2)
```

Note: The `anova()` command computes analysis of variance (or deviance) tables. When given one model as an argument, it displays the ANOVA table. When two (or more) nested models are given, it calculates the differences between them.

7.8.2   Log-likelihood

Example: 6.6.6

See 7.8.3 (AIC).

```
mod1 = lm(y ~ x1 + ... + xk, data=ds)
logLik(mod1)
```

Note: The `logLik()` can operate on `glm`, `lm`, `multinom`, `nls`, `Arima`, `gls`, `lme`, and `nlme` objects, among others.

7.8.3   Akaike Information Criterion (AIC)

Example: 6.6.6

See 7.8.2 (log-likelihood).

```
mod1 = lm(y ~ x1 + ... + xk, data=ds)
AIC(mod1)
```

Note: The `AIC()` function includes support for `glm`, `lm`, `multinom`, `nls`, `Arima`, `gls`, `lme`, and `nlme` objects.

7.8.4   Bayesian Information Criterion (BIC)

See 7.8.3 (AIC).

```
mod1 = lm(y ~ x1 + ... + xk, data=ds)
library(nlme)
BIC(mod1)
```

7.8.5   LASSO model

The LASSO (least absolute shrinkage and selection operator) is a model selection method for linear regression that minimizes the sum of squared errors subject to a constraint on the sum of the absolute value of the coefficients. This technique, due to Tibshirani [169], is particularly useful in data mining situations where a large number of predictors is being considered for inclusion in the model (see also 7.3.3).

```
library(lars)
lars(y ~ x1 + ... + xk, data=ds, type="lasso")
```

Note: The `lars()` function also implements least angle regression and forward stagewise methods.

7.8.6 Hosmer–Lemeshow goodness of fit

```
hosmerlem = function(y, yhat, g=10) {
    cutyhat = cut(yhat,
        breaks = quantile(yhat, probs=seq(0,
        1, 1/g)), include.lowest=TRUE)
    obs = xtabs(cbind(1 - y, y) ~ cutyhat)
    expect = xtabs(cbind(1 - yhat, yhat) ~ cutyhat)
    chisq = sum((obs - expect)^2/expect)
    P = 1 - pchisq(chisq, g-2)
    return(list(chisq=chisq, p.value=P))
}
```

Note: The test is straightforward to code directly. The `hosmerlem()` function accepts a vector of observed 0 and 1 outcomes and predicted probabilities. For a more refined version that accepts a model object as input, see `http://tinyurl.com/sasrblog-hosmer-lemeshow`.

7.8.7 Goodness of fit for count models

Example: 7.10.2

```
library(vcd)
poisfit = goodfit(x, "poisson")
```

The `goodfit()` function carries out a Pearson's χ^2 test of observed vs. expected counts. Other distributions supported include `binomial` and `nbinomial`.

Using the code below, R can also create a hanging rootogram [174] to assess the goodness of fit for count models. If the model fits well, then the bottom of each bar in the rootogram should be near zero.

```
library(vcd)
rootogram(poisfit)
```

7.9 Further resources

Many of the topics covered in this chapter are active areas of statistical research and many foundational articles are still useful. Here we provide references to texts that serve as accessible references.

Dobson and Barnett [32] is an accessible introduction to generalized linear models, while [111] remains a classic. Agresti [3] describes the analysis of categorical data. The CRAN statistics for the social sciences task view provides an overview of support in this area.

Fitzmaurice, Laird, and Ware [40] is an accessible overview of mixed effects methods while [185] reviews these methods for a variety of statistical packages. A comprehensive review of the material in this chapter is incorporated in [37]. The text by Hardin and Hilbe [59] provides a review of generalized estimating equations. The CRAN analysis of spatial data task view provides a summary of tools to read, visualize, and analyze spatial data.

Collett [26] is an accessible introduction to survival analysis. Manly [109] and Tabachnick and Fidell [164] provide a comprehensive introduction to multivariate statistics. Särndal, Swensson, and Wretman [148] provides a readable overview of the analysis of data from complex surveys.

7.10   Examples

To help illustrate the tools presented in this chapter, we apply many of the entries to the HELP data. The code can be downloaded from http://www.amherst.edu/~nhorton/r2/examples.

```
> options(digits=3)
> options(show.signif.stars=FALSE)
> load("savedfile")   # saved from previous chapter
```

The R dataset can be read in from a previously saved file (see p. 26 and 2.6.1).

7.10.1   Logistic regression

In this example we fit a logistic regression (7.1.1) modeling the probability of being homeless (spending one or more nights in a shelter or on the street in the past six months) as a function of predictors.

We use the glm() command to fit the logistic regression model.

```
> logres = glm(homeless ~ female + i1 + substance + sexrisk + indtot,
     binomial, data=ds)
> summary(logres)

Call:
glm(formula = homeless ~ female + i1 + substance + sexrisk +
    indtot, family = binomial, data = ds)

Deviance Residuals:
   Min      1Q  Median      3Q     Max
 -1.75   -1.04   -0.70    1.13    2.03

Coefficients:
                  Estimate Std. Error z value Pr(>|z|)
(Intercept)       -2.13192    0.63347   -3.37  0.00076
female            -0.26170    0.25146   -1.04  0.29800
i1                 0.01749    0.00631    2.77  0.00556
substancecocaine  -0.50335    0.26453   -1.90  0.05707
substanceheroin   -0.44314    0.27030   -1.64  0.10113
sexrisk            0.07251    0.03878    1.87  0.06152
indtot             0.04669    0.01622    2.88  0.00399

(Dispersion parameter for binomial family taken to be 1)

    Null deviance: 625.28  on 452  degrees of freedom
Residual deviance: 576.65  on 446  degrees of freedom
```

```
AIC: 590.7

Number of Fisher Scoring iterations: 4
```

There are a number of useful objects that be generated using the `summary()` function.

```
> names(summary(logres))
 [1] "call"           "terms"          "family"          "deviance"
 [5] "aic"            "contrasts"      "df.residual"     "null.deviance"
 [9] "df.null"        "iter"           "deviance.resid"  "coefficients"
[13] "aliased"        "dispersion"     "df"              "cov.unscaled"
[17] "cov.scaled"
> summary(logres)$coefficients
                  Estimate Std. Error z value Pr(>|z|)
(Intercept)       -2.1319    0.63347   -3.37 0.000764
female            -0.2617    0.25146   -1.04 0.297998
i1                 0.0175    0.00631    2.77 0.005563
substancecocaine  -0.5033    0.26453   -1.90 0.057068
substanceheroin   -0.4431    0.27030   -1.64 0.101128
sexrisk            0.0725    0.03878    1.87 0.061518
indtot             0.0467    0.01622    2.88 0.003993
```

7.10.2   Poisson regression

In this example we fit a Poisson regression model (7.1.6) for `i1`, the average number of drinks per day in the 30 days prior to entering the detox center.

```
> poisres = glm(i1 ~ female + substance + age, poisson, data=ds)
> summary(poisres)

Call:
glm(formula = i1 ~ female + substance + age, family = poisson,
    data = ds)

Deviance Residuals:
   Min      1Q  Median      3Q     Max
 -7.57   -3.69   -1.40    1.04   15.99


Coefficients:
                  Estimate Std. Error z value Pr(>|z|)
(Intercept)        2.89785    0.05827   49.73  < 2e-16
female            -0.17605    0.02802   -6.28  3.3e-10
substancecocaine  -0.81715    0.02776  -29.43  < 2e-16
substanceheroin   -1.12117    0.03392  -33.06  < 2e-16
age                0.01321    0.00145    9.08  < 2e-16

(Dispersion parameter for poisson family taken to be 1)

    Null deviance: 8898.9  on 452  degrees of freedom
Residual deviance: 6713.9  on 448  degrees of freedom
```

```
AIC: 8425

Number of Fisher Scoring iterations: 6
```

It is always important to check assumptions for models. This is particularly true for Poisson models, which are quite sensitive to model departures. There is support in the vcd package for a Pearson's χ^2 goodness-of-fit test.

```
> library(vcd)

Loading required package:  grid

> poisfit = with(ds, goodfit(e2b, "poisson"))
> summary(poisfit)

 Goodness-of-fit test for poisson distribution

                     X^2 df P(> X^2)
Likelihood Ratio 208 10   3.6e-39
```

The results indicate that the fit is poor ($\chi^2_{10} = 208$, $p < 0.0001$); the Poisson model does not appear to be tenable.

7.10.3 Zero-inflated Poisson regression

A zero-inflated Poisson regression model (7.2.1) might fit better. We'll allow a different probability of extra zeros per level of female.

```
> library(pscl)
```

```
> res = zeroinfl(i1 ~ female + substance + age | female, data=ds)
> res

Call:
zeroinfl(formula = i1 ~ female + substance + age | female, data = ds)

Count model coefficients (poisson with log link):
    (Intercept)           female  substancecocaine    substanceheroin
        3.05781          -0.06797          -0.72466           -0.76086
            age
        0.00927

Zero-inflation model coefficients (binomial with logit link):
(Intercept)         female
     -1.979          0.843
```

Women are more likely to abstain from alcohol than men: they have more than double the odds of being in the zero-inflation group ($p=0.0025$), and a smaller Poisson mean among those in the Poisson distribution ($p=0.015$). Other significant predictors include substance and age, though model assumptions for count models should always be carefully verified [69].

7.10.4   Negative binomial regression

A negative binomial regression model (7.1.7) might also improve on the Poisson.

```
> library(MASS)
> nbres = glm.nb(i1 ~ female + substance + age, data=ds)
> summary(nbres)

Call:
glm.nb(formula = i1 ~ female + substance + age, data = ds,
    init.theta = 0.810015139, link = log)

Deviance Residuals:
   Min      1Q  Median      3Q     Max
-2.414  -1.032  -0.278   0.241   2.808

Coefficients:
                  Estimate Std. Error z value Pr(>|z|)
(Intercept)        3.01693    0.28928   10.43  < 2e-16
female            -0.26887    0.12758   -2.11    0.035
substancecocaine  -0.82360    0.12904   -6.38  1.7e-10
substanceheroin   -1.14879    0.13882   -8.28  < 2e-16
age                0.01072    0.00725    1.48    0.139

(Dispersion parameter for Negative Binomial(0.81) family taken to be 1)

    Null deviance: 637.82  on 452  degrees of freedom
Residual deviance: 539.60  on 448  degrees of freedom
AIC: 3428

Number of Fisher Scoring iterations: 1


            Theta:  0.8100
        Std. Err.:  0.0589


 2 x log-likelihood:  -3416.3340
```

7.10.5   Quantile regression

In this section, we fit a quantile regression model (7.3.1) of the number of drinks (i1) as a function of predictors, modeling the 75th percentile (Q3).

```
> library(quantreg)
> quantres = rq(i1 ~ female + substance + age, tau=0.75, data=ds)

Warning:  Solution may be nonunique

> summary(quantres)

Warning:  Solution may be nonunique
```

```
Call: rq(formula = i1 ~ female + substance + age, tau = 0.75, data = ds)

tau: [1] 0.75

Coefficients:
                  coefficients lower bd upper bd
(Intercept)            29.636    14.150   42.603
female                 -2.909    -7.116    3.419
substancecocaine      -20.091   -29.011  -15.460
substanceheroin       -22.636   -28.256  -19.115
age                     0.182    -0.153    0.468
> detach(package:quantreg)
```

Because the quantreg package overrides needed functionality in other packages, we detach()
it after running the rq() function (see A.4.6).

7.10.6   Ordered logistic

To demonstrate an ordinal logit analysis (7.1.4), we first create an ordinal categorical variable from the sexrisk variable, then model this three-level ordinal variable as a function
of cesd and pcs.

```
> library(MASS)
> ds = mutate(ds, sexriskcat =
    as.factor(as.numeric(sexrisk >= 2) +
    as.numeric(sexrisk >= 6)))
> ologit = polr(sexriskcat ~ cesd + pcs, data=ds)
> summary(ologit)


Re-fitting to get Hessian

Call:
polr(formula = sexriskcat ~ cesd + pcs, data = ds)

Coefficients:
        Value Std. Error  t value
cesd -3.72e-05    0.00761 -0.00489
pcs   5.23e-03    0.00876  0.59649

Intercepts:
    Value  Std. Error t value
0|1 -1.669  0.562      -2.971
1|2  0.944  0.556       1.698

Residual Deviance: 871.76
AIC: 879.76
```

7.10.7   Generalized logistic model

We can fit a generalized logistic (7.1.5) model for the categorized sexrisk variable.

```
> library(VGAM)
> mlogit = vglm(sexriskcat ~ cesd + pcs,
    family=multinomial(refLevel=1), data=ds)
```

```
> summary(mlogit)

Call:
vglm(formula = sexriskcat ~ cesd + pcs, family = multinomial(refLevel = 1),
    data = ds)

Pearson residuals:
                     Min   1Q Median   3Q Max
log(mu[,2]/mu[,1])    -2 -0.6    0.8  0.8   1
log(mu[,3]/mu[,1])    -2 -0.4   -0.4  1.3   1

Coefficients:
              Estimate Std. Error z value
(Intercept):1    1.478       0.89     1.7
(Intercept):2    0.686       0.95     0.7
cesd:1          -0.013       0.01    -1.1
cesd:2          -0.007       0.01    -0.5
pcs:1            0.009       0.01     0.6
pcs:2            0.010       0.01     0.7

Number of linear predictors:  2

Names of linear predictors: log(mu[,2]/mu[,1]), log(mu[,3]/mu[,1])

Dispersion Parameter for multinomial family:    1

Residual deviance: 870 on 900 degrees of freedom

Log-likelihood: -435 on 900 degrees of freedom

Number of iterations: 5
> detach(package:VGAM)
```

Because the VGAM package overrides needed functionality in other packages, we detach() it after running the vglm() function (see A.4.6).

7.10.8 Generalized additive model

We can fit a generalized additive model (7.2.3), which we will later plot.

```
> library(gam)

Loaded gam 1.09.1

> gamreg= gam(cesd ~ female + lo(pcs) + substance, data=ds)
```

```
> summary(gamreg)

Call: gam(formula = cesd ~ female + lo(pcs) + substance, data = ds)
Deviance Residuals:
    Min      1Q  Median      3Q     Max
 -29.158  -8.136   0.811   8.226  29.250

(Dispersion Parameter for gaussian family taken to be 135)

    Null Deviance: 70788 on 452 degrees of freedom
Residual Deviance: 60288 on 445 degrees of freedom
AIC: 3519

Number of Local Scoring Iterations: 2

Anova for Parametric Effects
           Df Sum Sq Mean Sq F value  Pr(>F)
female      1   2202    2202    16.2 6.5e-05
lo(pcs)     1   5099    5099    37.6 1.9e-09
substance   2   1437     718     5.3  0.0053
Residuals 445  60288     135

Anova for Nonparametric Effects
            Npar Df Npar F Pr(F)
(Intercept)
female
lo(pcs)         3.1   3.77  0.01
substance
> coefficients(gamreg)
    (Intercept)            female          lo(pcs) substancecocaine
         46.524             4.339           -0.277           -3.956
  substanceheroin
         -0.205
```

The gam package provides a plot() method to display the results. The estimated smoothing function is provided in Figure 7.1.

7.10.9   Reshaping a dataset for longitudinal regression

A wide (multivariate) dataset can be reshaped (2.3.7) into a tall (longitudinal) dataset. Here we create time-varying variables (with a suffix tv) as well as keep baseline values (without the suffix).

```
> long = reshape(ds, idvar="id",
    varying=list(c("cesd1", "cesd2", "cesd3", "cesd4"),
              c("mcs1", "mcs2", "mcs3", "mcs4"),
              c("i11", "i12", "i13", "i14"),
              c("g1b1", "g1b2", "g1b3", "g1b4"),
              c("pcs1", "pcs2", "pcs3", "pcs4")),
    v.names=c("cesdtv", "mcstv", "i1tv", "g1btv", "pcstv"),
    timevar="time", times=1:4, direction="long")
```

```
> plot(gamreg, terms=c("lo(pcs)"), se=2, lwd=3)
> abline(h=0)
```

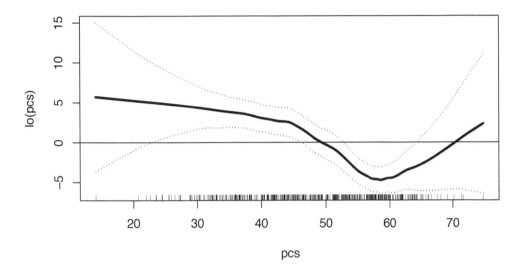

Figure 7.1: Scatterplots of smoothed association of physical component score (PCS) with CESD

While more complicated in this example (because of the need to keep the baseline values as non-time-varying variables), the `dplyr` and `tidyr` packages can also be used to accomplish the same task.

```
> library(dplyr); library(tidyr); library(stringr)
> long = gather(ds, measurement, value, cesd1, cesd2, cesd3, cesd4,
        pcs1, pcs2, pcs3, pcs4, mcs1, mcs2, mcs3, mcs4,
        i11, i12, i13, i14, g1b1, g1b2, g1b3, g1b4) %>%
  mutate(measurement = sub("(cesd|pcs|mcs|i1|g1b)([1234])",
    "\\1tv\\2", measurement)) %>%
  select(measurement, value, id, cesd, homeless, female, treat) %>%
  separate(measurement, into=c("measurement", "time"), sep=-2) %>%
  spread(measurement, value) %>%
  mutate(time = as.numeric(time)) %>%
  arrange(id, time)
```

We begin by gathering the variables by time, denote `measurement` as the new variable name, and `value` as the value. The `mutate()` function is used to modify the variable name (to indicate that it is time-varying). This requires use of a somewhat more complicated regular expression (Section 2.2.12). The `select()` function is called to bring in time-stationary variables before the `separate()` function creates a time indicator. The `spread()` function puts the variables into a single row per id and observation, `mutate()` turns the time variable into an integer, and `arrange()` sorts by id and time. These functions use the pipe operator to connect the operations (see A.5.3).

To check the results, we can compare the two datasets for the first subject (using the `filter()` function in the `dplyr` package).

```
> ds %>%
    filter(id==1) %>%
    select(id, pcs, pcs1, pcs2, pcs3, pcs4, mcs1, female, treat)
  id pcs pcs1 pcs2 pcs3 pcs4 mcs1 female treat
1  1 58.4 54.2   NA 52.1 52.3 52.2      0     1
> long %>%
    filter(id==1) %>%
    select(id, time, pcs, pcstv, mcstv, female, treat)
  id time  pcs pcstv mcstv female treat
1  1    1    1 58.4  54.2  52.2      0     1
2  1    2 58.4    NA    NA      0     1
3  1    3 58.4  52.1  56.1      0     1
4  1    4 58.4  52.3  58.0      0     1
```

Now that the dataset has been created, we can display the distribution of suicidal ideation over time.

```
> library(mosaic)
> tally(~ g1btv + time, data=long)
      time
g1btv    1   2   3   4
    0  219 187 225 245
    1   27  22  22  21
 <NA> 207 244 206 187
```

7.10.10   Linear model for correlated data

Here we fit a general linear model for correlated data (modeling the covariance matrix directly, 7.4.1).

```
> library(nlme)
> glsres = gls(cesdtv ~ treat + as.factor(time),
      correlation=corSymm(form = ~ time | id),
      weights=varIdent(form = ~ 1 | time),
      na.action=na.omit, data=long)
```

```
> summary(glsres)
Generalized least squares fit by REML
  Model: cesdtv ~ treat + as.factor(time)
  Data: long
   AIC  BIC logLik
  7550 7623  -3760

Correlation Structure: General
 Formula: ~time | id
 Parameter estimate(s):
 Correlation:
   1   2   3
2 0.584
```

```
3 0.639 0.743
4 0.474 0.585 0.735
Variance function:
 Structure: Different standard deviations per stratum
 Formula: ~1 | time
 Parameter estimates:
     1      3      4      2
 1.000  0.996  0.996  1.033

Coefficients:
                   Value Std.Error t-value p-value
(Intercept)        23.66     1.098   21.55   0.000
treat              -0.48     1.320   -0.36   0.716
as.factor(time)2    0.28     0.941    0.30   0.763
as.factor(time)3   -0.66     0.841   -0.78   0.433
as.factor(time)4   -2.41     0.959   -2.52   0.012

 Correlation:
                   (Intr) treat  as.()2 as.()3
treat             -0.627
as.factor(time)2  -0.395  0.016
as.factor(time)3  -0.433  0.014  0.630
as.factor(time)4  -0.464  0.002  0.536  0.708

Standardized residuals:
    Min      Q1     Med      Q3     Max
 -1.643  -0.874  -0.115   0.708   2.582

Residual standard error: 14.4
Degrees of freedom: 969 total; 964 residual
```

```
> anova(glsres)
Denom. DF: 964
                 numDF F-value p-value
(Intercept)          1    1168  <.0001
treat                1       0  0.6887
as.factor(time)      3       4  0.0145
```

A set of side-by-side boxplots (8.2.2) by time can be generated using the following commands (see Figure 7.2).

7.10.11 Linear mixed (random slope) model

Here we fix a mixed-effects, or random slope model (7.4.3). In this example, we specify a categorical fixed effect of time but a random slope across time treated continuously. We do this by making a copy of the time variable in a new dataset. First we create an `as.factor()` version of time. As an alternative, we could nest the call to `as.factor()` within the call to `lme()`.

```
> library(lattice)
> bwplot(cesdtv ~ as.factor(treat)| time, xlab="TREAT",
      strip=strip.custom(strip.names=TRUE, strip.levels=TRUE),
      ylab="CESD", layout=c(4,1), col="black", data=long,
      par.settings=list(box.rectangle=list(col="black"),
         box.dot=list(col="black"), box.umbrella=list(col="black"))))
```

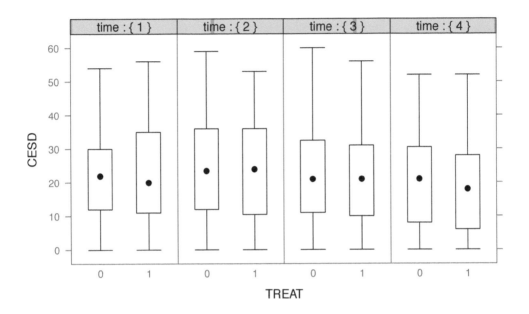

Figure 7.2: Side-by-side box plots of CESD by treatment and time

```
> long = transform(long, tf=as.factor(time))
> library(nlme)
> lmeslope = lme(fixed=cesdtv ~ treat + tf,
      random= ~ time | id, na.action=na.omit,
      data=long)
> print(lmeslope)
Linear mixed-effects model fit by REML
  Data: long
  Log-restricted-likelihood: -3772
  Fixed: cesdtv ~ treat + tf
(Intercept)      treat         tf2        tf3        tf4
    23.8843    -0.4353     -0.0615    -1.0142    -2.5776

Random effects:
 Formula: ~time | id
 Structure: General positive-definite, Log-Cholesky parametrization
           StdDev Corr
(Intercept) 13.73  (Intr)
time         3.03  -0.527
```

```
Residual      7.85

Number of Observations: 969
Number of Groups: 383
```

```
> anova(lmeslope)
            numDF denDF F-value p-value
(Intercept)     1   583    1163  <.0001
treat           1   381       0  0.7257
tf              3   583       3  0.0189
```

We use the `random.effects` and `predict()` functions to find the predicted random effects and predicted values, respectively.

```
> reffs = random.effects(lmeslope)
> reffs[1,]
  (Intercept)    time
1       -13.5  -0.024
```

```
> predval = predict(lmeslope, newdata=long, level=0:1)
> predval[predval$id==1,]
    id predict.fixed predict.id
1.1  1          23.4       9.94
1.2  1          23.4       9.86
1.3  1          22.4       8.88
1.4  1          20.9       7.30
```

```
> vc = VarCorr(lmeslope)
> summary(vc)
   Variance    StdDev      Corr
    9.17:1     3.03:1        :1
   61.58:1     7.85:1   -0.527:1
  188.43:1    13.73:1   (Intr):1
```

The `VarCorr()` function calculates the variances, standard deviations, and correlations between the random effects terms, as well as the within-group error variance and standard deviation.

7.10.12 Generalized estimating equations

We fit a GEE model (7.4.7), using an exchangeable working correlation matrix and empirical variance [99].

```
> library(gee)
> sortlong = long[order(long$id),]
> geeres = gee(formula = g1btv ~ treat + time, id=id, data=sortlong,
      family=binomial, na.action=na.omit, corstr="exchangeable")
```

```
Beginning Cgee S-function, @(#) geeformula.q 4.13 98/01/27
running glm to get initial regression estimate

(Intercept)        treat          time
   -1.9649        0.0443        -0.1256
```

In addition to returning an object with results, the `gee()` function displays the coefficients from a model assuming that all observations are uncorrelated.

```
> coef(geeres)
(Intercept)        treat          time
   -1.85169      -0.00874      -0.14593
> sqrt(diag(geeres$robust.variance))
(Intercept)        treat          time
    0.2723        0.2683        0.0872
> geeres$working.correlation
       [,1]  [,2]  [,3]  [,4]
[1,] 1.000 0.299 0.299 0.299
[2,] 0.299 1.000 0.299 0.299
[3,] 0.299 0.299 1.000 0.299
[4,] 0.299 0.299 0.299 1.000
```

7.10.13 Generalized linear mixed model

Here we fit a GLMM (7.4.6), predicting recent suicidal ideation as a function of treatment and time. Each subject is assumed to have their own random intercept.

```
> library(lme4)
> glmmres = glmer(g1btv ~ treat + time + (1|id),
    family=binomial(link="logit"), data=long)
```

```
> summary(glmmres)
Generalized linear mixed model fit by maximum likelihood (Laplace
  Approximation) [glmerMod]
 Family: binomial  ( logit )
Formula: g1btv ~ treat + time + (1 | id)
   Data: long

     AIC       BIC    logLik deviance df.resid
     509       528      -250      501      964

Scaled residuals:
    Min      1Q Median      3Q     Max
-2.0990 -0.0251 -0.0183 -0.0152  2.4879

Random effects:
 Groups Name         Variance Std.Dev.
 id     (Intercept)  61.2     7.82
Number of obs: 968, groups:  id, 383
```

```
Fixed effects:
            Estimate Std. Error z value Pr(>|z|)
(Intercept)  -6.8699     0.8151   -8.43   <2e-16
treat        -0.0537     0.6986   -0.08    0.939
time         -0.3556     0.1724   -2.06    0.039

Correlation of Fixed Effects:
      (Intr) treat
treat -0.431
time  -0.362  0.002
```

7.10.14   Cox proportional hazards model

Here we fit a proportional hazards model (7.5.1) for the time to linkage to primary care, with randomization group, age, gender, and CESD as predictors.

We request the Efron estimator (default), which provides a better approximation in the case of many ties.

```
> library(survival)
> survobj = coxph(Surv(dayslink, linkstatus) ~ treat + age + female +
     cesd, method="efron", data=ds)
> print(survobj)
Call:
coxph(formula = Surv(dayslink, linkstatus) ~ treat + age + female +
    cesd, data = ds, method = "efron")

           coef exp(coef) se(coef)      z     p
treat   1.65509     5.234  0.19324  8.565 0.000
age     0.02474     1.025  0.01032  2.397 0.017
female -0.32569     0.722  0.20382 -1.598 0.110
cesd    0.00237     1.002  0.00638  0.372 0.710

Likelihood ratio test=95  on 4 df, p=0  n= 431, number of events= 163
    (22 observations deleted due to missingness)
```

7.10.15   Cronbach's α

We calculate Cronbach's α (7.6.1) for the 20 items comprising the CESD (Center for Epidemiologic Studies–Depression scale).

```
> library(multilevel)
> with(ds, cronbach(cbind(f1a, f1b, f1c, f1d, f1e, f1f, f1g, f1h,
    f1i, f1j, f1k, f1l, f1m, f1n, f1o, f1p, f1q, f1r,
    f1s, f1t)))
$Alpha
[1] 0.761
```

```
$N
[1] 446
```

The observed α of 0.76 from the HELP study is relatively low: this may be due to ceiling effects for this sample of subjects recruited in a detoxification unit.

7.10.16   Factor analysis

Here we consider a maximum likelihood factor analysis (7.6.2) with varimax rotation for the individual items of the CESD (Center for Epidemiologic Studies–Depression) scale. The individual questions can be found in Table B.2. We arbitrarily force three factors. Before beginning, we exclude observations with missing values.

```
> res = with(ds, factanal(~ f1a + f1b + f1c + f1d + f1e + f1f + f1g + f1h +
    f1i + f1j + f1k + f1l + f1m + f1n + f1o + f1p + f1q + f1r +
    f1s + f1t, factors=3, rotation="varimax", na.action=na.omit,
    scores="regression"))
> print(res, cutoff=0.45, sort=TRUE)

Call:
factanal(x = ~f1a + f1b + f1c + f1d + f1e + f1f + f1g + f1h +
  f1i + f1j + f1k + f1l + f1m + f1n + f1o + f1p + f1q + f1r +
  f1s + f1t, factors = 3, na.action = na.omit, scores = "regression",
  rotation = "varimax")

Uniquenesses:
  f1a   f1b   f1c   f1d   f1e   f1f   f1g   f1h   f1i   f1j   f1k   f1l
0.745 0.768 0.484 0.707 0.701 0.421 0.765 0.601 0.616 0.625 0.705 0.514
  f1m   f1n   f1o   f1p   f1q   f1r   f1s   f1t
0.882 0.623 0.644 0.407 0.713 0.467 0.273 0.527

Loadings:
    Factor1 Factor2 Factor3
f1c  0.618
f1e  0.518
f1f  0.666
f1k  0.523
f1r  0.614
f1h         -0.621
f1l         -0.640
f1p         -0.755
f1o                  0.532
f1s                  0.802
f1a
f1b
f1d         -0.454
f1g  0.471
f1i  0.463
f1j  0.495
f1m
f1n  0.485
```

```
f1q  0.457
f1t  0.489


             Factor1 Factor2 Factor3
SS loadings    3.847   2.329   1.636
Proportion Var 0.192   0.116   0.082
Cumulative Var 0.192   0.309   0.391


Test of the hypothesis that 3 factors are sufficient.
The chi square statistic is 289 on 133 degrees of freedom.
The p-value is 1.56e-13
```

Next, we interpret the item scores from the output. We see that the second factor loads on the reverse coded items (H, L, P, and D, see 2.6.3). Factor 3 loads on items O and S (*people were unfriendly* and *I felt that people dislike me*).

7.10.17 Recursive partitioning

In this example, we use recursive partitioning (7.6.3) to classify subjects based on their homeless status, using gender, drinking, primary substance, RAB sexrisk, MCS, and PCS as predictors.

```
> library(rpart)
> ds = transform(ds, sub = as.factor(substance))
> homeless.rpart = rpart(homeless ~ female + i1 + sub + sexrisk + mcs +
    pcs, method="class", data=ds)
> printcp(homeless.rpart)

Classification tree:
rpart(formula = homeless ~ female + i1 + sub + sexrisk + mcs +
    pcs, data = ds, method = "class")

Variables actually used in tree construction:
[1] female  i1      mcs     pcs     sexrisk

Root node error: 209/453 = 0.5

n= 453

      CP nsplit rel error xerror xstd
1 0.10      0       1.0    1.0 0.05
2 0.05      1       0.9    1.0 0.05
3 0.03      4       0.8    1.0 0.05
4 0.02      5       0.7    1.0 0.05
5 0.01      7       0.7    0.9 0.05
6 0.01      9       0.7    0.9 0.05
```

```
> library(partykit)
> plot(as.party(homeless.rpart))
```

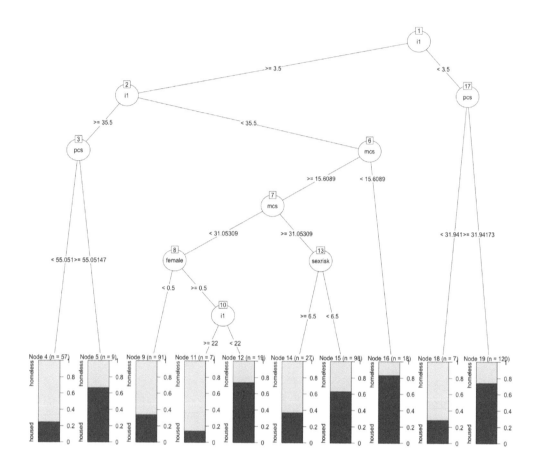

Figure 7.3: Recursive partitioning tree

Figure 7.3 displays the tree. To help interpret this model, we can calculate the empirical proportion of subjects that were homeless among those with $i1 < 3.5$ by `pcs` less than 31.94 (this corresponds to Nodes 18 and 19 in the figure).

```
> home = with(ds, homeless[i1<3.5])
> pcslow = with(ds, pcs[i1<3.5]<=31.94)
> table(home, pcslow)
    pcslow
home FALSE TRUE
   0    89    2
   1    31    5
> rm(home, pcslow)
```

Among this subset, 71.4% (5 of 7) of those with low PCS scores are homeless, while only 25.8% (31 of 120) of those with PCS scores above the threshold are homeless.

7.10.18   Linear discriminant analysis

We use linear discriminant analysis (7.6.4) to distinguish between homeless and nonhomeless subjects.

```
> library(MASS)
> ngroups = length(unique(ds$homeless))
> ldamodel = lda(homeless ~ age + cesd + mcs + pcs,
    prior=rep(1/ngroups, ngroups), data=ds)
```

```
> print(ldamodel)
Call:
lda(homeless ~ age + cesd + mcs + pcs, data = ds, prior = rep(1/ngroups,
    ngroups))

Prior probabilities of groups:
  0   1
0.5 0.5

Group means:
   age cesd  mcs  pcs
0 35.0 31.8 32.5 49.0
1 36.4 34.0 30.7 46.9

Coefficients of linear discriminants:
         LD1
age   0.0702
cesd  0.0269
mcs  -0.0195
pcs  -0.0426
```

The results indicate that homeless subjects tend to be older, have higher CESD scores, and lower MCS and PCS scores. Figure 7.4 displays the distribution of linear discriminant function values by homeless status; the discrimination ability appears to be slight. The distribution of the linear discriminant function values is shifted to the right for the homeless subjects, though there is considerable overlap between the groups. Details on the display of `lda` objects can be found using `help(plot.lda)`.

7.10.19   Hierarchical clustering

In this example, we use hierarchical clustering (7.6.6) to group continuous variables from the HELP dataset.

```
> cormat = with(ds, cor(cbind(mcs, pcs, cesd, i1, sexrisk),
    use="pairwise.complete.obs"))
> hclustobj = hclust(dist(cormat))
```

Figure 7.5 displays the clustering. Not surprisingly, the MCS and PCS variables cluster together, since they both utilize similar questions and structures. The CESD and I1 variables cluster together, while there is a separate node for SEXRISK.

```
> plot(ldamodel)
```

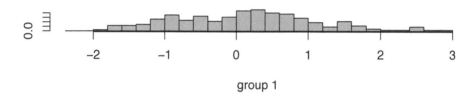

Figure 7.4: Graphical display of assignment probabilities or score functions from linear discriminant analysis by actual homeless status

```
> plot(hclustobj)
```

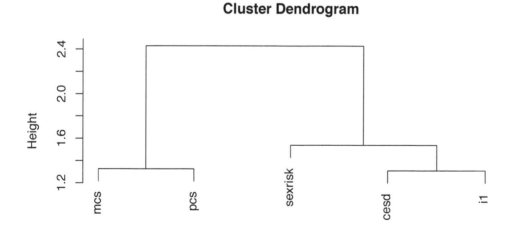

Figure 7.5: Results from hierarchical clustering

Chapter 8

A graphical compendium

This chapter provides a compendium of graphical displays. More details about configuration options can be found in Chapter 9. Examples appear throughout the book.

Producing graphics for data analysis is relatively simple. Producing graphics for publication is more complex and typically requires a great deal of time to achieve the desired appearance. Our intent is to provide sufficient guidance so that most effects can be achieved, but further investigation of the documentation and experimentation will doubtless be necessary for specific needs. There are a huge number of options: we aim to provide a roadmap as well as examples to illustrate the power of the package.

While many graphics can be generated using a single command, within the base graphics system, figures are often built up element by element. For example, an empty box can be created with a specific set of x and y axis labels and tick marks, then points can be added with different printing characters. Text annotations can then be added, along with legends and other additional information (see 6.6.1). The CRAN graphics task view (`http://cran.r-project.org/web/views`) provides a comprehensive listing of functionality to create graphics.

A somewhat intimidating number of options are available, some of which can be specified using the `par()` graphics parameters (see 9.2), while others can be given as options to plotting commands (such as `plot()` or `lines()`).

A number of graphics devices are available that support different platforms and formats. The default varies by platform (`Windows()` under Windows, `X11()` under Linux, and `quartz()` under modern Mac OS X distributions). A device is created automatically when a plotting command is run, or a device can be started in advance to create a file in a particular format (e.g., the `pdf()` device).

A series of powerful add-on packages to create sophisticated graphics is available. These include the `grid` package [118], the `lattice` package [147], and the `ggplot2` package [188]. We will focus primarily on base graphics, but include examples of the `lattice` and `ggplot2` systems.

Running `example()` for a specified function is often helpful for commands shown in this chapter, as is `demo(graphics)`.

8.1 Univariate plots

8.1.1 Barplot

While not typically an efficient graphical display, there are times when a barplot is appropriate to display counts by groups.

```
barplot(table(x1, x2), legend=c("grp1", "grp2"), xlab="X2")
```
or
```
library(lattice)
barchart(table(x1, x2, x3))
```

Note: The input for the `barplot()` function is given as a vector or matrix of bar heights, while the `barchart()` function within the `lattice` package supports three-dimensional tables (see `example(barplot)` and `example(barchart)`). A similar `dotchart()` function produces a horizontal slot for each group with a dot reflecting the frequency.

8.1.2   Stem-and-leaf plot

Example: 12.4.4

Stem-and-leaf plots are text-based graphics that are particularly useful to describe the distribution of small datasets. They are often used for teaching purposes.

```
stem(x)
```

Note: The `scale` option can be used to increase or decrease the number of stems (default value is 1).

8.1.3   Dotplot

Dotplots (also called Wilkinson dotplots) are a simple introductory graphic useful for presenting numerical data when the dataset is small [196]. They are often used for teaching purposes, as they help to identify clusters and gaps while conserving numerical information (in the same manner as a stem-and-leaf plot).

```
library(mosaic)
dotPlot(x)
```

8.1.4   Histogram

Example: 5.7.1

The example in 5.7.1 demonstrates how to annotate a histogram with an overlaid normal or kernel density estimate. Similar estimates are available for all other densities supported (see Table 3.1).

```
hist(x)
```

Note: The default behavior for a histogram is to display frequencies on the vertical axis; probability densities can be displayed using the `freq=FALSE` option. The default title is given by `paste("Histogram of" , x)` where x is the name of the variable being plotted; this can be changed with the `main` option. The `histogram()` function in the `lattice` package provides an alternative implementation.

8.1.5   Density plot

A density plot displays a nonparametric estimate of the empirical probability density function (see also 8.2.3, overlaid density plots, and 8.3.5, bivariate density plots).

```
plot(density(x))
```
or
```
library(lattice)
densityplot(~ x2, data=ds)
```
Note: The help page for `density()` provides guidance for specifying the default window width and other options for `densityplot()`.

8.1.6 Empirical cumulative probability density plot

```
plot(ecdf(x))
```

Note: The `knots()` function can be used to determine when the empirical density function jumps.

8.1.7 Boxplot

Examples: 6.6.6 and 7.10.10

See also 8.2.2 (side-by-side boxplots).

```
boxplot(x)
```

Note: The `boxplot()` function allows sideways orientation using the `horizontal=TRUE` option. The `lattice` package provides an alternative implementation using the `bwplot()` function.

8.1.8 Violin plots

Violin plots combine a boxplot and (doubled) kernel density plot.

```
library(vioplot)
vioplot(x2[x1==0], x2[x1==1])
```

Note: Here we assume that `x1` has two levels (0 and 1).

8.2 Univariate plots by grouping variable

8.2.1 Side-by-side histograms

```
library(lattice)
histogram(~ x2 | x1)
```

8.2.2 Side-by-side boxplots

See also 8.1.7 (boxplots)

```
boxplot(y[x==0], y[x==1], y[x==2], names=c("X=0", "X=1", "X=2"))
```
or
```
library(lattice)
bwplot(y ~ x)
```

Note: The `boxplot()` function can be given multiple arguments of vectors to display or can use a formula interface (which will generate a boxplot for each level of the variable x). A number of useful options are available, including `varwidth` to draw the boxplots with widths proportional to the square root of the number of observations in that group, `horizontal` to reverse the default orientation, `notch` to display notched boxplots, and `names` to specify a vector of labels for the groups. Boxplots can also be created using the `bwplot()` function in the `lattice` package.

8.2.3   Overlaid density plots

See also 8.1.5 (density plots) and 8.3.5 (bivariate density).

```
library(lattice)
densityplot(~ x2, groups=x1, auto.key=TRUE)
```

Note: This code shows the density for a single variable plotted for each level of a second variable.

8.2.4   Bar chart with error bars

While the graphical display with a bar, the height of which indicates the mean and vertical lines indicating the standard error is quite common, many find these displays troubling. We concur with graphics authorities such as Edward Tufte [172], who discourage their use, as does Frank Harrell's group at Vanderbilt (see `biostat.mc.vanderbilt.edu/wiki/Main/StatisticalPolicy`).

```
library(lattice)
library(grid)
dynamitePlot = function(height, error,
   names=as.character(1:length(height)),
   significance=NA, ylim=c(0, maxLim), ...)
{
   if (missing(error)) { error = 0 }
   maxLim = 1.2 * max(mapply(sum, height, error))
   mError = min(c(error, na.rm=TRUE))
   barchart(height ~ names, ylim=ylim, panel=function(x,y,...) {
      panel.barchart(x, y, ...)
      grid.polyline(c(x,x), c(y, y+error), id=rep(x,2),
         default.units='native',
         arrow=arrow(angle=45, length=unit(mError, 'native')))
      grid.polyline(c(x,x), c(y, y-error), id=rep(x,2),
         default.units='native',
         arrow=arrow(angle=45, length=unit(mError, 'native')))
      grid.text(x=x, y=y + error + .05*maxLim, label=significance,
         default.units='native')
   }, ...)
}
```

Note: This graph is built up in parts using customized calls to the `barchart()` function. Much of the code (due to Randall Pruim) involves setting up the appropriate axis limits, as a function of the heights and error ranges, then drawing the lines adding the text using calls to `grid.polyline()` and `grid.text()`. The ... option to the function passes any additional arguments to the `panel.barchart()` function, to allow further customization (see the book by Sarkar [147]). Once defined, the function can be run using the following syntax.

```
Values = c(1, 2, 5, 4)
Errors = c(0.25, 0.5, 0.33, 0.12)
Names = paste("Trial", 1:4)
Sig = c("a", "a", "b", "b")
dynamitePlot(Values, Errors, names=Names, significance=Sig)
```

The `bargraph.CI()` function within the `sciplot` package provides similar functionality.

8.3 Bivariate plots

8.3.1 Scatterplot

Example: 6.6.1

See 8.3.2 (scatterplot with multiple y values) and 8.4.1 (matrix of scatterplots).

```
plot(x, y)
```

Note: Many objects have default plotting methods (e.g., for a linear model object, `plot.lm()` is called). More information can be found using `methods(plot)`. Specifying `type="n"` causes nothing to be plotted (but sets up axes and draws boxes, see 3.4.1). This technique is often useful if a plot is built up part by part.

8.3.2 Scatterplot with multiple y values

Example: 8.7.1

See also 8.4.1 (matrix of scatterplots).

```
plot(x, y1, pch=pchval1)   # create 1 plot with single y axis
points(x, y2, pch=pchval2)
...
points(x, yk, pch=pchvalk)
```
or
```
# create 1 plot with 2 separate y axes
addsecondy = function(x, y, origy, yname="Y2") {
   prevlimits = range(origy)
   axislimits = range(y)
   axis(side=4, at=prevlimits[1] + diff(prevlimits)*c(0:5)/5,
      labels=round(axislimits[1] + diff(axislimits)*c(0:5)/5, 1))
   mtext(yname, side=4)
   newy = (y-axislimits[1])/(diff(axislimits)/diff(prevlimits)) +
      prevlimits[1]
   points(x, newy, pch=2)
}
plottwoy = function(x, y1, y2, xname="X", y1name="Y1", y2name="Y2")
{
   plot(x, y1, ylab=y1name, xlab=xname)
   addsecondy(x, y2, y1, yname=y2name)
}
plottwoy(x, y1, y2, y1name="Y1", y2name="Y2")
```

Note: To create a figure with a single y axis value, it is straightforward to repeatedly call `points()` or other functions to add elements.

In the second example, two functions `addsecondy()` and `plottwoy()` are defined to add points on a new scale and an appropriate axis on the right. This involves rescaling and labeling the second axis (`side=4`) with 6 tick marks, as well as rescaling the y2 variable.

8.3.3   Scatterplot with binning

When there are many observations, scatterplots can become difficult to read because many plot symbols will obscure one another. One solution to this problem is to use binned scatterplots. Other options are transparent plot symbols that display as darker areas when overlaid (8.3.4) and bivariate density plotting (8.3.5).

```
library(hexbin)
plot(hexbin(x, y))
```

8.3.4   Transparent overplotting scatterplot

When there are many observations, scatterplots can become difficult to read because many plot symbols will obscure one another. One solution to this problem is to use transparent plot symbols that display as darker areas when overlaid. Other options are binned scatterplots (8.3.3) and bivariate density plotting (8.3.5).

```
plot(x1, x2, pch=19, col="#00000022", cex=0.1)
```

8.3.5   Bivariate density plot

When there are many observations, scatterplots can become difficult to read because many plot symbols will obscure one another. One solution to this problem is bivariate density plots, which also have other uses. Other options are binned scatterplots (8.3.3) and transparent overplotting (8.3.4).

```
smoothScatter(x, y)
```
or
```
library(GenKern)
# bivariate density
op = KernSur(x, y, na.rm=TRUE)
image(op$xords, op$yords, op$zden, col=gray.colors(100), axes=TRUE,
    xlab="x var", ylab="y var")
```
Note: The `smoothScatter()` function provides a simple interface for a bivariate density plot. The default smoother for `KernSur()` can be specified using the `kernel` option (possible values include the default Gaussian, rectangular, triangular, Epanechnikov, biweight, cosine, or optcosine). Bivariate density support is provided with the `GenKern` package. Any of the three-dimensional plotting routines (see 8.4.4) can be used to visualize the results.

8.3.6 Scatterplot with marginal histograms

Example: 8.7.3

```
scatterhist = function(x, y, xlab="x label", ylab="y label"){
   zones=matrix(c(2,0,1,3), ncol=2, byrow=TRUE)
   layout(zones, widths=c(4/5,1/5), heights=c(1/5,4/5))
   xhist = hist(x, plot=FALSE)
   yhist = hist(y, plot=FALSE)
   top = max(c(xhist$counts, yhist$counts))
   par(mar=c(3,3,1,1))
   plot(x,y)
   par(mar=c(0,3,1,1))
   barplot(xhist$counts, axes=FALSE, ylim=c(0, top), space=0)
   par(mar=c(3,0,1,1))
   barplot(yhist$counts, axes=FALSE, xlim=c(0, top), space=0, horiz=TRUE)
   par(oma=c(3,3,0,0))
   mtext(xlab, side=1, line=1, outer=TRUE, adj=0,
      at=.8 * (mean(x) - min(x))/(max(x)-min(x)))
   mtext(ylab, side=2, line=1, outer=TRUE, adj=0,
      at=(.8 * (mean(y) - min(y))/(max(y) - min(y))))
}
scatterhist(x, y)
```

Note: In this entry we demonstrate how to build a more complicated figure in pieces using base graphics. The `layout()` function splits the graphics region into four non-equal parts, then individual plotting functions are called.

8.4 Multivariate plots

8.4.1 Matrix of scatterplots

Example: 8.7.6

```
pairs(data.frame(x1, ..., xk))
```

Note: The `pairs()` function is quite flexible, since it calls user-specified functions to determine what to display on the lower triangle, diagonal, and upper triangle (`examples(pairs)` illustrate its capabilities). The `ggpairs()` function in the `GGally` package can be used to create a pairs plot with both continuous and categorical variables.

8.4.2 Conditioning plot

A conditioning plot is used to display a scatter plot for each level of one or two classification variables.

Example: 8.7.2

```
coplot(y ~ x1 | x2*x3)
```

Note: The `coplot()` function displays plots of y and x1, stratified by x2 and x3. All variables may be either numeric or factors.

8.4.3   Contour plots

A contour plot shows the Cartesian plain with similar valued points linked by lines. The most familiar versions of such plots may be contour maps showing lines of constant elevations that are very useful for hiking.

```
contour(x, y, z)
```
or
```
filled.contour(x, y, z)
```

Note: The `contour()` function displays a standard contour plot. The `filled.contour()` function creates a contour plot with colored areas between the contours.

8.4.4   3-D plots

Perspective or surface plots and needle plots can be used to visualize data in three dimensions. These are particularly useful when a response is observed over a grid of two-dimensional values.

```
persp(x, y, z)
image(x, y, z)

library(scatterplot3d)
scatterplot3d(x, y, z)
```

Note: The values provided for x and y must be in ascending order.

8.5   Special-purpose plots

8.5.1   Choropleth maps

Example: 12.3.3

```
library(ggmap)
mymap = map_data('state')   # need to add variable to plot
p0 = ggplot(map_data, aes(x=x, y=y, group=z)) +
    geom_polygon(aes(fill = cut_number(z, 5))) +
    geom_path(colour = 'gray', linestyle = 2) +
    scale_fill_brewer(palette = 'PuRd') +
    coord_map();
plot(p0)
```

Note: More examples of maps can be found in the `ggmap` package documentation.

8.5.2   Interaction plots

Example: 6.6.6

Interaction plots are used to display means by two variables (as in a two-way analysis of variance, 6.1.9).

```
interaction.plot(x1, x2, y)
```

Note: The default statistic to compute is the mean; other options can be specified using the `fun` option.

8.5.3 Plots for categorical data

A variety of less-traditional plots can be used to graphically represent categorical data. While these tend to have a low data-to-ink ratio, they can be useful in figures with repeated multiples [171].

```
mosaicplot(table(x, y, z))
assocplot(table(x, y))
```

Note: The `mosaicplot()` function provides a graphical representation of a two-dimensional or higher contingency table, with the area of each box representing the number of observations in that cell. The `assocplot()` function can be used to display the deviations from independence for a two-dimensional contingency table. Positive deviations of observed minus expected counts are above the line and colored black, while negative deviations are below the line and colored red. Tables can be included in graphics using the `gridExtra` packages (see 5.7.3).

8.5.4 Circular plot

Circular plots are used to analyze data that wraps (e.g., directions expressed as angles, time of day on a 24-hour clock) [39, 82].

```
library(circular)
plot.circular(x, stack=TRUE, bins=50)
```

8.5.5 Plot an arbitrary function

Example: 10.1.6

```
curve(expr, from=start, to=stop, n=number)
```
or
```
x = seq(from=start, to=stop, by=step)
y = expr(x)
plot(x, y)
```

Note: The `curve()` function can be used to plot an arbitrary function denoted by `expr`, with `n` values of `x` between `start` and `stop` (see 9.1.5). This can also be built up in parts by use of the `seq()` function. The `plotFun()` function in the `mosaic` package can be used to plot regression models or other functions generated using `makeFun()`.

8.5.6 Normal quantile–quantile plot

Example: 6.6.4

Quantile–quantile plots are a commonly used graphical technique to assess whether a univariate sample of random variables is consistent with a Gaussian (normal) distribution.

```
qqnorm(x)
qqline(x)
```

Note: The `qqline()` function adds a straight line that goes through the first and third quartiles.

8.5.7   Receiver operating characteristic (ROC) curve

Example: 8.7.5

See also 5.2.2 (diagnostic agreement) and 7.1.1 (logistic regression).

Receiver operating characteristic curves can be used to help determine the optimal cut-score to predict a dichotomous measure. This is particularly useful in assessing diagnostic accuracy in terms of sensitivity (the probability of detecting the disorder if it is present), specificity (the probability that a disorder is not detected if it is not present), and the area under the curve (AUC). The variable x represents a predictor (e.g., individual scores) and y a dichotomous outcome. There is a close connection between the idea of the ROC curve and goodness of fit for logistic regression, where the latter allows multiple predictors to be used. Support is provided within the ROCR package [160].

```
library(ROCR)
pred = prediction(x, y)
perf = performance(pred, "tpr", "fpr")
plot(perf)
```

Note: The area AUC can be calculated by specifying "auc" as an argument when calling the performance() function.

8.5.8   Plot confidence intervals for the mean

```
pred.w.clim = predict(lm(y ~ x), interval="confidence")
matplot(x, pred.w.clim, lty=c(1, 2, 2), type="l", ylab="predicted y")
```
or
```
library(mosaic)
xyplot(y ~ x, panel=panel.lmbands)
```

Note: The first entry produces fit and confidence limits at the original observations in the original order. If the observations aren't sorted relative to the explanatory variable x, the resulting plot will be a jumble. The matplot() function is used to generate lines, with a solid line (lty=1) for predicted values and dashed line (lty=2) for the confidence bounds. The panel.lmbands() function in the mosaic package can also be used to plot these values.

8.5.9   Plot prediction limits from a simple linear regression

```
pred.w.plim = predict(lm(y ~ x), interval="prediction")
matplot(x, pred.w.plim, lty=c(1, 2, 2), type="l", ylab="predicted y")
```

Note: This entry produces fit and confidence limits at the original observations in the original order. If the observations aren't sorted relative to the explanatory variable x, the resulting plot will be a jumble. The matplot() function is used to generate lines, with a solid line (lty=1) for predicted values and dashed line (lty=2) for the confidence bounds.

8.5.10   Plot predicted lines for each value of a variable

Here we describe how to generate plots for a variable X_1 versus Y separately for each value of the variable X_2 (see conditioning plot, 8.4.2).

```
plot(x1, y, pch=" ") # create an empty plot of the correct size
abline(lm(y ~ x1, subset=x2==0), lty=1, lwd=2)
abline(lm(y ~ x1, subset=x2==1), lty=2, lwd=2)
...
abline(lm(y ~ x1, subset=x2==k), lty=k+1, lwd=2)
```

Note: The `abline()` function is used to generate lines for each of the subsets, with a solid line (`lty=1`) for the first group and a dashed line (`lty=2`) for the second (this assumes that X_2 takes on values 0–k, see 11.1.2). The `plotFun()` function in the `mosaic` package provides another way of adding lines or arbitrary curves to a plot.

8.5.11   Kaplan–Meier plot

Example: 8.7.4

See also 5.4.6 (log-rank test).

```
library(survival)
fit = survfit(Surv(time, status) ~ as.factor(x), data=ds)
plot(fit, conf.int=FALSE, lty=1:length(unique(x)))
```

Note: The `Surv()` function is used to combine survival time and status, where `time` is length of follow-up (interval censored data can be accommodated via an additional parameter) and `status=1` indicates an event (e.g., death) while `status=0` indicates censoring. The model is stratified by each level of the group variable `x` (see adding legends, 9.1.15, and different line styles, 9.2.11). More information can be found in the CRAN survival analysis task view.

8.5.12   Hazard function plotting

The hazard function from data censored on the right can be estimated using kernel methods (see also 8.7.4).

```
library(muhaz)
plot(muhaz(time, status))
```

Note: Estimation of hazards from multiple groups can be plotted together using the `lines()` command after running `muhaz()` on the other groups.

8.5.13   Mean–difference plots

The Tukey mean–difference plot, popularized in medical research as the Bland–Altman plot [6], plots the difference between two variables against their mean. This can be useful when they are two different methods for assessing the same quantity.

```
baplot = function(x, y) {
   bamean = (x + y)/2
   badiff = (y - x)
   plot(badiff ~ bamean, pch=20, xlab="mean", ylab="difference")
   abline(h = c(mean(badiff), mean(badiff)+1.96 * sd(badiff),
      mean(badiff)-1.96 * sd(badiff)), lty=2)
}
baplot(x, y)
```

8.6   Further resources

The books by Tufte [170, 171, 172, 173] provide an excellent framework for graphical displays, some of which build on the work of Tukey [174]. Comprehensive and accessible books on graphics include [118], [147], and [188] (see also the help files at `http://docs.ggplot2.org/current`).

8.7   Examples

To help illustrate the tools presented in this chapter, we apply many of the entries to the HELP data. The code can be downloaded from `http://www.amherst.edu/~nhorton/r2/examples`. We begin by reading in the data.

```
> options(digits=3)
> ds = read.csv("http://www.amherst.edu/~nhorton/r2/datasets/help.csv")
```

8.7.1   Scatterplot with multiple axes

The following example creates a single figure that displays the relationship between CESD and the variables `indtot` (Inventory of Drug Abuse Consequences, InDUC) and `mcs` (Mental Component Score) for a subset of female alcohol-involved subjects. We specify two different y-axes (8.3.2) for the figure. Some housekeeping is needed. The second y variable must be rescaled to the range of the original, and the axis labels and tick marks added on the right. To accomplish this, we write a function `plottwoy()`, which first makes the plot of the first (left axis) y against x, adds a lowess curve through that data, then calls a second function, `addsecondy()`.

```
> plottwoy = function(x, y1, y2, xname="X", y1name="Y1", y2name="Y2") {
    plot(x, y1, ylab=y1name, xlab=xname)
    lines(lowess(x, y1), lwd=3)
    addsecondy(x, y2, y1, yname=y2name)
  }
```

The function `addsecondy()` does the work of rescaling the range of the second variable to that of the first, adds the right axis, and plots a lowess curve through the data for the rescaled `y2` variable.

```
> addsecondy = function(x, y, origy, yname="Y2") {
    prevlimits = range(origy)
    axislimits = range(y)
    axis(side=4, at=prevlimits[1] + diff(prevlimits)*c(0:5)/5,
        labels=round(axislimits[1] + diff(axislimits)*c(0:5)/5, 1))
    mtext(yname, side=4)
    newy = (y-axislimits[1])/(diff(axislimits)/diff(prevlimits)) +
        prevlimits[1]
    points(x, newy, pch=2)
    lines(lowess(x, newy), lty=2, lwd=3)
  }
```

Finally, the newly defined functions can be run and Figure 8.1 generated.

```
> with(ds, plottwoy(cesd[female==1&substance=="alcohol"],
    indtot[female==1&substance=="alcohol"],
    mcs[female==1&substance=="alcohol"], xname="cesd",
    y1name="indtot", y2name="mcs"))
```

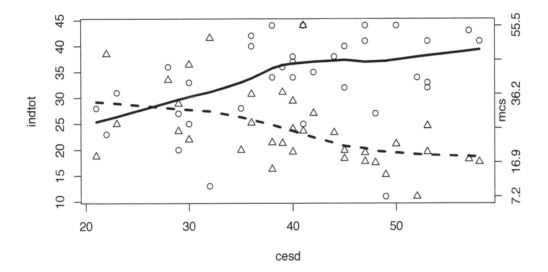

Figure 8.1: Plot of InDUC and MCS vs. CESD for female alcohol-involved subjects

8.7.2 Conditioning plot

Figure 8.2 displays a conditioning plot (8.4.2) with the association between MCS and CESD stratified by substance and report of suicidal thoughts (g1b). We first ensure that the necessary packages are loaded (A.6.1).

```
> library(lattice)
```

Then we can set up and generate the plot. There is a similar association between CESD and MCS for each of the substance groups. Subjects with suicidal thoughts tended to have higher CESD scores, and the association between CESD and MCS was somewhat less pronounced than for those without suicidal thoughts.

The lattice package has a number of settings that can be controlled by the user. We have specified the mosaic black-and-white theme. Figure 8.3 displays the configuration chosen for the display of Figure 8.2.

8.7.3 Scatterplot with marginal histograms

We can assess the univariate as well as bivariate distribution of the MCS and CESD scores using a scatterplot with a marginal histogram (8.3.6), as shown in Figure 8.4.
We use the layout() function (9.2.3) to create the graphic.

```
> trellis.par.set(theme=col.mosaic(bw=TRUE))
> ds = transform(ds, suicidal.thoughts = ifelse(g1b==1, "Y", "N"))
> coplot(mcs ~ cesd | suicidal.thoughts*substance,
    panel=panel.smooth, data=ds)
```

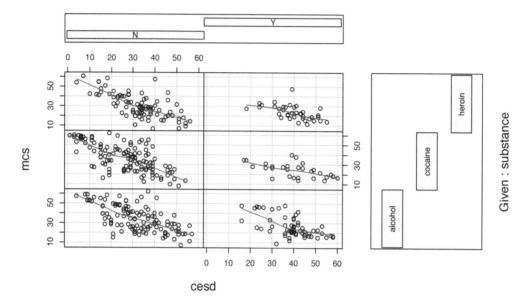

Figure 8.2: Association of MCS and CESD, stratified by substance and report of suicidal thoughts

```
> scatterhist = function(x, y, xlab="x label", ylab="y label", cexval=1.3){
    zones=matrix(c(3,1,2,4), ncol=2, byrow=TRUE)
    layout(zones, widths=c(4/5,1/5), heights=c(1/5,4/5))
    par(mar=c(0,0,0,0))
    plot(type="n",x=1, y =1, bty="n",xaxt="n", yaxt="n")
    text(x=1,y=1,paste0("n = ",min(length(x), length(y))), cex=cexval)
    xhist = hist(x, plot=FALSE)
    yhist = hist(y, plot=FALSE)
    top = max(c(xhist$counts, yhist$counts))
    par(mar=c(3.9,3.9,1,1))
    plot(x,y, xlab=xlab, ylab=ylab, cex.sub=cexval,
        pch=19, col="#00000044")
    lines(lowess(x, y), lwd=2)
    par(mar=c(0,3,1,1))
    barplot(xhist$counts, axes=FALSE, ylim=c(0, top), space=0)
    par(mar=c(3,0,1,1))
    barplot(yhist$counts, axes=FALSE, xlim=c(0, top), space=0, horiz=TRUE)
    par(oma=c(3,3,0,0))
  }
```

```
> show.settings()
```

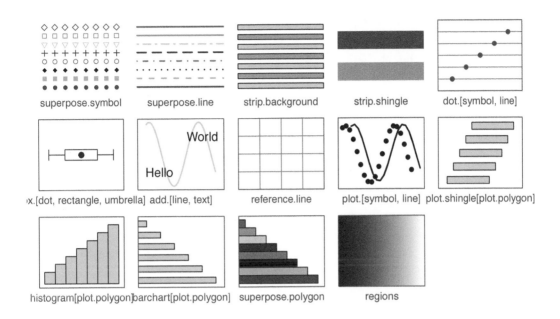

Figure 8.3: Lattice settings using the mosaic black-and-white theme

8.7.4 Kaplan–Meier plot

The main outcome of the HELP study was time to linkage to primary care, as a function of randomization group. This can be displayed using a Kaplan–Meier plot (see 8.5.11). Detailed information regarding the Kaplan–Meier estimator at each time point can be found by calling summary(survobj). Figure 8.5 displays the estimates, with + signs indicating censored observations.

```
> library(survival)
> survobj = survfit(Surv(dayslink, linkstatus) ~ treat, data=ds)
> print(survobj)
Call: survfit(formula = Surv(dayslink, linkstatus) ~ treat, data = ds)

   22 observations deleted due to missingness
        records n.max n.start events median 0.95LCL 0.95UCL
treat=0     209   209     209     35     NA      NA      NA
treat=1     222   222     222    128    120      79     272
```

As reported previously [72, 145], there is a highly statistically significant effect of treatment, with approximately 55% of clinic subjects linking to primary care, as opposed to only 15% of control subjects.

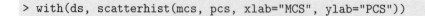

```
> with(ds, scatterhist(mcs, pcs, xlab="MCS", ylab="PCS"))
```

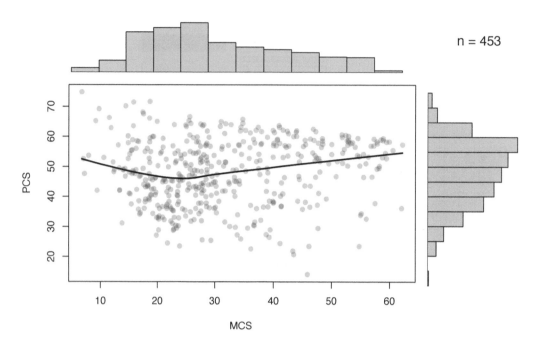

Figure 8.4: Association of MCS and PCS with marginal histograms

8.7.5   ROC curve

Receiver operating characteristic (ROC) curves are used for diagnostic agreement (5.2.2 and 8.5.7) as well as assessing goodness of fit for logistic regression (7.1.1). These are easily created using the ROCR package. Figure 8.6 displays the receiver operating characteristic curve predicting suicidal thoughts using the CESD measure of depressive symptoms.

We first load the ROCR package, create a prediction object, and retrieve the area under the curve (AUC) to use in Figure 8.6.

```
> library(ROCR)
> pred = with(ds, prediction(cesd, g1b))
> auc = slot(performance(pred, "auc"), "y.values")[[1]]
```

We can then plot the ROC curve, adding a display of cutoffs for particular CESD values ranging from 20 to 50. These values are offset from the ROC curve using the text.adj option.

If the continuous variable (in this case cesd) is replaced by the predicted probability from a logistic regression model, multiple predictors can be included.

8.7.6   Pairs plot

We can qualitatively assess the associations between some of the continuous measures of mental health, physical health, and alcohol consumption using a pairs plot or scatterplot matrix (8.4.1). To make the results clearer, we include only the female subjects.

```
> plot(survobj, lty=1:2, lwd=2, col=c(4,2))
> title("Product-Limit Survival Estimates")
> legend(20, .38, legend=c("Control", "Treatment"), lty=c(1,2), lwd=2,
      col=c(4,2), cex=1.2)
```

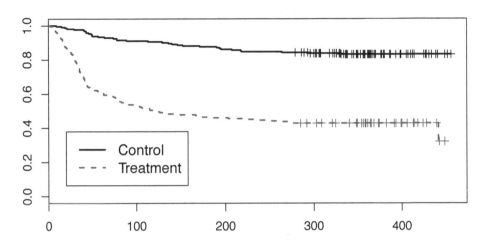

Figure 8.5: Kaplan–Meier estimate of time to linkage to primary care by randomization group

A simple version with only the scatterplots could be generated easily with the **pairs()** function (results not shown):

```
> pairs(c(ds[72:74], ds[67]))
```

or

```
> pairs(ds[c("pcs", "mcs", "cesd", "i1")])
```

Here instead, we demonstrate building a figure using several functions. We begin with a function **panel.hist()** to display the diagonal entries (in this case, by displaying a histogram).

```
> panel.hist = function(x, ...)
  {
    usr = par("usr"); on.exit(par(usr))
    par(usr=c(usr[1:2], 0, 1.5))
    h = hist(x, plot=FALSE)
    breaks = h$breaks; nB = length(breaks)
    y = h$counts; y = y/max(y)
```

```
> plot(performance(pred, "tpr", "fpr"),
    print.cutoffs.at=seq(from=20, to=50, by=5),
    text.adj=c(1, -.5), lwd=2)
> lines(c(0, 1), c(0, 1))
> text(.6, .2, paste("AUC=", round(auc,3), sep=""), cex=1.4)
> title("ROC Curve for Model")
```

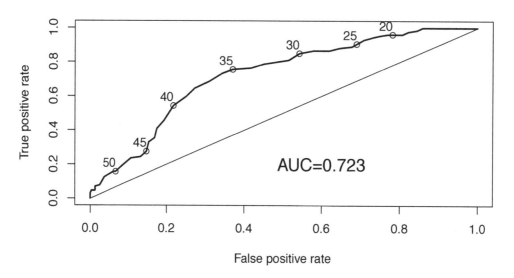

Figure 8.6: Receiver operating characteristic curve for the logistical regression model predicting suicidal thoughts using the CESD as a measure of depressive symptoms (sensitivity = true positive rate; 1-specificity = false positive rate)

```
    rect(breaks[-nB], 0, breaks[-1], y, col="cyan", ...)
}
```

Another function is created to create a scatterplot along with a fitted line.

```
> panel.lm = function(x, y, col=par("col"), bg=NA, pch=par("pch"),
    cex=1, col.lm="red", ...)
{
    points(x, y, pch=pch, col=col, bg=bg, cex=cex)
    ok = is.finite(x) & is.finite(y)
    if (any(ok))
        abline(lm(y[ok] ~ x[ok]))
}
```

These functions are called (along with the built-in panel.smooth() function) to display the results. Figure 8.7 displays the pairs plot of CESD, MCS, PCS, and I1, with histograms along the diagonals. For R, smoothing splines are fit on the lower triangle, linear fits on the upper triangle, using code fragments derived from example(pairs).

```
> pairs(~ cesd + mcs + pcs + i1, subset=(female==1),
    lower.panel=panel.smooth, diag.panel=panel.hist,
    upper.panel=panel.lm, data=ds)
```

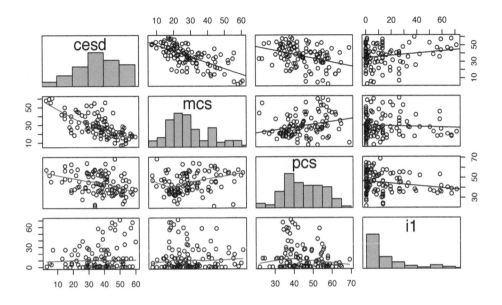

Figure 8.7: Pairs plot of variables from the HELP dataset using the `lattice` package

There is an indication that CESD, MCS, and PCS are interrelated, while I1 appears to have modest associations with the other variables.

The GGally package can be used to create pairs plots for the alcohol-involved subjects with a mixture of categorical and continuous variables (see Figure 8.8).

We begin by subsetting the data, making a version of the `female` variable with text values, and then selecting only a few variables.

```
> library(GGally)
> library(dplyr)
> smallds = ds %>%
    filter(substance=="alcohol") %>%
    mutate(sex = ifelse(female==1, "female", "male")) %>%
    select(cesd, mcs, sex)
```

8.7.7 Visualize correlation matrix

One visual analysis that might be helpful to display would be the pairwise correlations. We utilize the approach used by Sarkar to re-create Figure 13.5 of the *Lattice: Multivariate Data Visualization with R* book [147]. Other examples in that reference help to motivate the power of the `lattice` package far beyond what is provided by `demo(lattice)`.

```
> ggpairs(smallds,
    axisLabels="show",
    diag = list(continuous = "bar", discrete = "bar"),
    upper = list(continuous = "points", combo = "box"),
    lower = list(continuous = "cor", combo = "facethist"))
```

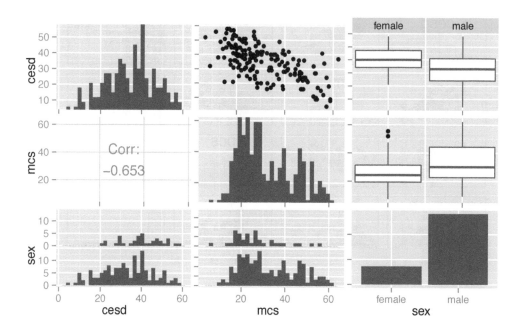

Figure 8.8: Pairs plot of variables from the HELP dataset using the GGally package.

```
> cormat = with(ds, cor(cbind(mcs, pcs, pss_fr, drugrisk,
    cesd, indtot, i1, sexrisk), use="pairwise.complete.obs"))
> oldopt = options(digits=2)
> cormat
           mcs    pcs pss_fr drugrisk   cesd indtot     i1 sexrisk
mcs      1.000  0.110  0.138  -0.2058 -0.682  -0.38 -0.087 -0.1061
pcs      0.110  1.000  0.077  -0.1411 -0.293  -0.13 -0.196  0.0239
pss_fr   0.138  0.077  1.000  -0.0390 -0.184  -0.20 -0.070 -0.1128
drugrisk -0.206 -0.141 -0.039  1.0000  0.179   0.18 -0.100 -0.0055
cesd    -0.682 -0.293 -0.184   0.1789  1.000   0.34  0.176  0.0157
indtot  -0.381 -0.135 -0.198   0.1807  0.336   1.00  0.202  0.1132
i1      -0.087 -0.196 -0.070  -0.0999  0.176   0.20  1.000  0.0881
sexrisk -0.106  0.024 -0.113  -0.0055  0.016   0.11  0.088  1.0000
> options(oldopt)
```

```
> ds$drugrisk[is.na(ds$drugrisk)] = 0
> panel.corrgram = function(x, y, z, at, level=0.9, label=FALSE, ...)
  {
    require(ellipse, quietly=TRUE)
    zcol = level.colors(z, at=at, col.regions=gray.colors)
```

```
  for (i in seq(along=z)) {
    ell = ellipse(z[i], level=level, npoints=50,
      scale=c(.2, .2), centre=c(x[i], y[i]))
    panel.polygon(ell, col=zcol[i], border=zcol[i], ...)
  }
  if (label)
    panel.text(x=x, y=y, lab=100*round(z, 2), cex=0.8,
      col=ifelse(z < 0, "white", "black"))
}
```

```
> library(ellipse); library(lattice)
> levelplot(cormat, at=do.breaks(c(-1.01, 1.01), 20), xlab=NULL, ylab=NULL,
    colorkey=list(space = "top", col=gray.colors),
    scales=list(x=list(rot=90)), panel=panel.corrgram, labels=TRUE)
```

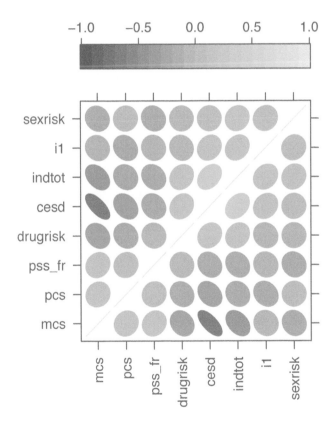

Figure 8.9: Visual display of correlations (times 100)

Chapter 9

Graphical options and configuration

This chapter describes how to annotate graphical displays and change defaults to create publication-quality figures, as well as details regarding how to output graphics in a variety of file formats (9.3).

9.1 Adding elements

It is relatively simple to add features to graphs that have been generated by one of the functions discussed in Chapter 8 that uses base graphics. For example, adding an arbitrary line to a graphic requires only one function call with the two endpoints as arguments (9.1.1). Other mechanisms to add to existing graphs created using the **lattice** package are available (see the `layer()` function).

9.1.1 Arbitrary straight line

Example: 6.6.1

```
plot(x, y)
lines(point1, point2)
```
or
```
abline(intercept, slope)
```
Note: The `lines()` function draws lines between successive pairs of locations specified by `point1` and `point2`, which are each vectors with values for the x and y axes. The `abline()` function draws a line based on the slope-intercept form. Vertical or horizontal lines can be specified using the `v` or `h` option to `abline()`.

9.1.2 Plot symbols

Example: 5.7.2

```
plot(x, y, pch=pchval)
```
or
```
points(x, y, string, pch=pchval)
```
or

```
library(lattice)
xyplot(x ~ y, group=factor(groupvar), data=ds)
```
or
```
library(ggplot2)
qplot(x, y, col=factor(groupvar), shape=factor(groupvar), data=ds)
```

Note: The `pch` option requires either a single character or an integer code. Some useful values include 20 (dot), 46 (point), 3 (plus), 5 (diamond), and 2 (triangle) (running `example(points)` will display more possibilities). The size of the plotting symbol can be changed using the `cex` option. The vector function `text()` adds the value in the variable `string` to the plot at the specified location. The examples using `xyplot()` and `qplot()` will also generate scatterplots with different plot symbols for each level of `groupvar`.

9.1.3 Add points to an existing graphic

Example: 6.6.1

See also 9.1.2 (specifying plotting character).

```
plot(x, y)
points(x, y)
```

9.1.4 Jitter points

Example: 5.7.2

Jittering is the process of adding a negligible amount of noise to each observed value so that the number of observations sharing a value can be easily discerned.

```
jitterx = jitter(x)
```

Note: The default value for the range of the random uniforms is 40% of the smallest difference between values.

9.1.5 Regression line fit to points

```
plot(x, y)
abline(lm(y ~ x))
```

Note: The `abline()` function accepts regression objects with a single predictor as input.

9.1.6 Smoothed line

Example: 5.7.2

See also 7.10.8 (generalized additive models).

```
plot(...)
lines(lowess(x, y))
```

Note: The `f` parameter to `lowess()` can be specified to control the proportion of points that influence the local value (larger values give more smoothness). The `supsmu()` (Friedman's "super smoother") and `loess()` (local polynomial regression fitting) functions are alternative smoothers.

9.1.7 Normal density

Example: 6.6.4

A normal density plot can be added as an annotation to a histogram or empirical density.

```
hist(x)
xvals = seq(from=min(x), to=max(x), length=100)
lines(pnorm(xvals, mean(x), sd(x)))
```

9.1.8 Marginal rug plot

Example: 5.7.2

A rug plot displays the marginal distribution on one of the margins of a scatterplot.

```
rug(x, side=sideval)
```

Note: The `rug()` function adds a marginal plot to one of the sides of an existing plot (`sideval=1` for bottom (default), 2 for left, 3 for top, and 4 for right side).

9.1.9 Titles

Example: 5.7.4

```
title(main="main", sub="sub", xlab="xlab", ylab="ylab")
```

Note: The title commands refer to the main title, subtitle, x axis, and y axis, respectively. Some plotting commands (e.g., `hist()`) create titles by default, and the appropriate option within those routines needs to be specified when calling them.

9.1.10 Footnotes

```
title(sub="sub")
```

Note: The `sub` option for the `title()` function generates a subtitle.

9.1.11 Text

Example: 5.7.2, 8.7.5

```
text(x, y, labels)
```

Note: Each value of the character vector `labels` is displayed at the specified (X,Y) coordinate. The `adj` option can be used to change text justification to the left, center (default), or right of the coordinate. The `srt` option can be used to rotate text, while `cex` controls the size of the text. The `font` option to `par()` allows specification of plain, bold, italic, or bold italic fonts (see the `family` option to specify the name of a different font family).

9.1.12   Mathematical symbols

Example: 3.4.1

```
plot(x, y)
text(x, y, expression(mathexpression))
```

Note: The `expression()` argument can be used to embed mathematical expressions and symbols (e.g., $\mu = 0$, $\sigma^2 = 4$) in graphical displays as text, axis labels, legends, or titles. See `help(plotmath)` for more details on the form of `mathexpression` and `example(plotmath)` for examples.

9.1.13   Arrows and shapes

Example: 5.7.4 and 8.7.6

```
arrows(x1, y1, x2, y2)
rect(x1, y1, x2, y2)
polygon(x, y)

library(plotrix)
draw.circle(x, y, r)
```

Note: The `arrows()`, `rect()`, and `polygon()` functions take vectors as arguments and create the appropriate object with vertices specified by each element of those vectors.

9.1.14   Add grid

A rectangular grid can sometimes be helpful to add to an existing plot.

```
grid(nx=num, ny=num)
```

Note: The `nx` and `ny` options control the number of cells in the grid. If they are specified as `NULL`, the grid aligns with the tick marks. The default shading is light gray, with a dotted line. Further support for complex grids is available within the `grid.lines()` function in the `grid` package.

9.1.15   Legend

Example: 3.4.1 and 5.7.4

```
plot(x, y)
legend(xval, yval, legend=c("Grp1","Grp2"), lty=1:2, col=3:4)
```

Note: The `legend()` command can be used to add a legend at the location (`xval`, `yval`) to distinguish groups on a display. Line styles (9.2.11) and colors (9.2.13) can be used to distinguish the groups. A vector of legend labels, line types, and colors can be specified using the `legend`, `lty`, and `col` options, respectively.

9.1.16   Identifying and locating points

```
locator(n)
```

Note: The `locator()` function identifies the position of the cursor when the mouse button is pressed. An optional argument n specifies how many values to return. The `identify()` function works in the same fashion, but returns the point closest to the cursor.

9.2 Options and parameters

Many options can be given to plots generated using base graphics. These are generally arguments to `plot()`, `par()`, or other high-level functions (see the documentation for the `par()` function).

9.2.1 Graph size

```
pdf("filename.pdf", width=Xin, height=Yin)
```

Note: The graph size is specified as an optional argument when starting a graphics device (e.g., `pdf()`, 9.3.1), with arguments Xin and Yin given in inches. Other devices have similar arguments to specify the size.

9.2.2 Grid of plots per page

Example: 6.6.4

See also 9.2.3 (more general page layouts).

```
par(mfrow=c(a, b))
```
or
```
par(mfcol=c(a, b))
```

Note: The `mfrow` option specifies that plots will be drawn in an a × b array by row (by column for `mfcol`).

9.2.3 More general page layouts

Example: 8.7.3

See also 9.2.2 (grid of plots per page).

```
oldpar = par(no.readonly=TRUE)
layout(mat, widths=wvec, heights=hvec)
layout.show()
#fill the layout with plots
par(oldpar)
```

Note: The `layout()` function divides the graphics device into rows and columns, the relative sizes of which are specified by the `widths` and `heights` vectors. The number of rows and columns, plus the locations in the matrix to which the figures will be plotted, are specified by `mat`. The elements of the `mat` matrix are integers showing the order in which generated plots fill the cells. Larger and smaller figures can be plotted by repeating some integer values. The `layout.show(n)` function plots the outline of the next n figures. Other options to arrange plots on a device include `par(mfrow)`, for regular grids of plots, and `split.screen()`. The `lattice` package provides a different implementation of multiple plots (see the `position` option).

9.2.4   Fonts

```
pdf(file="plot.pdf")
par(family="Palatino")
plot(x, y)
dev.off()
```

Note: Supported postscript families include `AvantGarde`, `Bookman`, `Courier`, `Helvetica`, `Helvetica-Narrow`, `NewCenturySchoolbook`, `Palatino`, and `Times` (see `?postscript`).

9.2.5   Point and text size

Example: 6.6.6

```
plot(x, y, cex=cexval)
```

Note: The `cex` option specifies how much the plotting text and symbols should be magnified relative to the default value of 1 (see `help(par)` for details on how to specify this for axes, labels, and titles, e.g., `cex.axis`).

9.2.6   Box around plots

Example: 5.7.4

```
plot(x, y, bty=btyval)
```

Note: Control for the box around the plot can be specified using `btyval`, where if the argument is one of `o` (the default), `1`, `7`, `c`, `u`, or `]`, the resulting box resembles the corresponding character, while a value of `n` suppresses the box.

9.2.7   Size of margins

Example: 6.6.4

Margin options control how tight plots are to the printable area.

```
par(mar=c(bot, left, top, right),    # inner margin
    oma=c(bot, left, top, right))    # outer margin
```

Note: The vector given to `mar` specifies the number of lines of margin around a plot: the default is `c(5, 4, 4, 2) + 0.1`. The `oma` option specifies additional lines outside the entire plotting area (the default is `c(0,0,0,0)`). Other options to control margin spacing include `omd` and `omi`.

9.2.8   Graphical settings

Example: 6.6.4

```
# change values, while saving old
oldvalues = par(...)

# restore old values for graphical settings
par(oldvalues)
```

9.2.9 Axis range and style

Example: 8.7.3

```
plot(x, y, xlim=c(minx, maxx), ylim=c(miny, maxy), xaxs="i", yaxs="i")
```

Note: The `xaxs` and `yaxs` options control whether tick marks extend beyond the limits of the plotted observations (default) or are constrained to be internal (`"i"`). More control is available through the `axis()` and `mtext()` functions.

9.2.10 Axis labels, values, and tick marks

Example: 3.4.1

```
plot(x, y, lab=c(x, y, len),  # number of tick marks
   las=lasval,    # orientation of tick marks
   tck=tckval,    # length of tick marks
   tcl=tclval,    # length of tick marks
   xaxp=c(x1, x2, n),   # coordinates of the extreme tick marks
   yaxp=c(x1, x2, n),   # coordinates of the extreme tick marks
   xlab="X axis label", ylab="Y axis label")
```

Note: Options for `las` include 0 for always parallel, 1 for always horizontal, 2 for perpendicular, and 3 for vertical.

9.2.11 Line styles

Example: 6.6.4

```
plot(...)
lines(x, y, lty=ltyval)
```

Note: Supported line type values include 0=blank, 1=solid (default), 2=dashed, 3=dotted, 4=dotdash, 5=longdash, and 6=twodash.

9.2.12 Line widths

Example: 3.4.1

```
plot(...)
lines(x, y, lwd=lwdval)
```

Note: The default for `lwd` is 1; the value of `lwdval` must be positive.

9.2.13 Colors

Example: 5.7.4

```
plot(...)
lines(x, y, col=colval)
```

Note: For more information on setting colors, see the `Color Specification` section within `help(par)` as well as `demo(colors)`. The `colors()` function lists available colors, while the `colors.plot()` function within the **epitools** package displays a matrix of colors, and `colors.matrix()` returns a matrix of color names. The `display.brewer.all()` function within the **RColorBrewer** package is particularly useful for selecting a set of complementary colors for a palette.

9.2.14   Log scale

```
plot(x, y, log=logval)
```

Note: A natural log scale can be specified using the `log` option to `plot()`, where `log="x"` denotes only the x axis, `"y"` only the y axis, and `"xy"` for both.

9.2.15   Omit axes

Example: 12.2

```
plot(x, y, xaxt="n", yaxt="n")
```

9.3   Saving graphs

It is straightforward to export graphics in a variety of formats. In addition to what is described below, RStudio allows export of a plot in multiple formats along with full control over the size and resolution.

9.3.1   PDF

```
pdf("file.pdf")
plot(...)
dev.off()
```

Note: The `dev.off()` function is used to close a graphics device.

9.3.2   Postscript

```
postscript("file.ps")
plot(...)
dev.off()
```

Note: The `dev.off()` function is used to close a graphics device.

9.3.3   RTF

Rich Text Format (RTF) is a file format developed for cross-platform document sharing. Many word processors are able to read and write RTF documents.

It's also possible to generate Microsoft Word files through use the `markdown` package and `Pandoc`. This process is simplified in RStudio (see 11.3).

```
library(rtf)
output = "file.doc"
rtf = RTF(output)
addParagraph(rtf, "This is a plot.\n")
addPlot(rtf, plot.fun=plot,width=6, height=6, res=300, x, y)
done(rtf)
```

Note: This example adds text and the equivalent of `plot(x, y)` to an RTF file. The `rtf` package vignette provides additional details for formatting graphs and tables.

9.3.4 JPEG

```
jpeg("filename.jpg")
plot(...)
dev.off()
```

Note: The `dev.off()` function is used to close a graphics device.

9.3.5 Windows Metafile

```
win.metafile("file.wmf")
plot(...)
dev.off()
```

Note: The function `win.metafile()` is only supported under Windows. Functions that generate multiple plots are not supported. The `dev.off()` function is used to close a graphics device.

9.3.6 Bitmap image file (BMP)

```
bmp("filename.bmp")
plot(...)
dev.off()
```

Note: The `dev.off()` function is used to close a graphics device.

9.3.7 Tagged Image File Format

```
tiff("filename.tiff")
plot(...)
dev.off()
```

Note: The `dev.off()` function is used to close a graphics device.

9.3.8 PNG

```
png("filename.png")
plot(...)
dev.off()
```

Note: The `dev.off()` function is used to close a graphics device.

9.3.9 Closing a graphic device

Example: 8.7.4

The following code closes a graphics window. This is particularly useful when a graphics file is being created.

```
dev.off()
```

Chapter 10

Simulation

Simulations provide a powerful way to answer questions and explore properties of statistical estimators and procedures. In this chapter, we will explore how to simulate data in a variety of common settings, and apply some of the techniques introduced earlier.

10.1 Generating data

10.1.1 Generate categorical data

Simulation of data from continuous probability distributions is straightforward using the functions detailed in 3.1.1. Simulating from categorical distributions can be done manually or using some available functions.

```
> options(digits=3)
> options(width=72) # narrow output
> p = c(.1,.2,.3)
> x = runif(10000)
> mycat1 = numeric(10000)
> for (i in 0:length(p)) {
      mycat1 = mycat1 + (x >= sum(p[0:i]))
      }
> table(mycat1)
mycat1
   1    2    3    4
 955 1988 3028 4029
```

```
> mycat2 = cut(runif(10000), c(0, 0.1, 0.3, 0.6, 1))
> summary(mycat2)
  (0,0.1] (0.1,0.3] (0.3,0.6]   (0.6,1]
     1050      2033      3041      3876
> mycat3 = sample(1:4, 10000, rep=TRUE, prob=c(.1,.2,.3,.4))
> table(mycat3)
mycat3
   1    2    3    4
1023 2015 3009 3953
```

The `cut()` function (2.2.4) bins continuous data into categories with both endpoints defined by the arguments. Note that the `min()` and `max()` functions can be particularly useful here in the outer categories. The `sample()` function as shown treats the values $1, 2, 3, 4$ as a dataset, and samples from the dataset 10,000 times with the probability of selection defined in the `prob` vector.

10.1.2   Generate data from a logistic regression

Here we show how to simulate data from a logistic regression (7.1.1). Our process is to generate the linear predictor, then apply the inverse link, and finally draw from a distribution with this parameter. This approach is useful in that it can easily be applied to other generalized linear models (7.1). Here we make the intercept -1, the slope 0.5, and generate $5,000$ observations.

```
> intercept = -1
> beta = 0.5
> n = 5000
> xtest = rnorm(n, mean=1, sd=1)
> linpred = intercept + (xtest * beta)
> prob = exp(linpred)/(1 + exp(linpred))
> ytest = ifelse(runif(n) < prob, 1, 0)
```

While the results of `summary()` for a `glm` object is relatively concise, we can display just the estimated values of the coefficients from the logistic regression model using the `coef()` function (see 6.4.1).

```
> coef(glm(ytest ~ xtest, family=binomial))
(Intercept)        xtest
    -1.005        0.483
```

10.1.3   Generate data from a generalized linear mixed model

In this example, we generate data from a generalized linear mixed model (7.4.6) with a dichotomous outcome. We generate 1500 clusters, denoted by `id`. There is one predictor with a common value for all observations in a cluster (X_1). Each observation within the cluster has an order indicator (denoted by X_2) that has a linear effect (`beta_2`), and there is an additional predictor that varies among observations (X_3). The dichotomous outcome Y is generated from these predictors using a logistic link incorporating a normal distributed random intercept for each cluster.

The simulation approach is an extension of that shown in the previous section (see also 4.1.3).

```
> n = 1500; p = 3; sigbsq = 4
> beta = c(-2, 1.5, 0.5, -1)
> id = rep(1:n, each=p)          # 1 1 ... 1 2 2 ... 2 ... n
> x1 = as.numeric(id < (n+1)/2) # 1 1 ... 1 0 0 ... 0
> randint = rep(rnorm(n, 0, sqrt(sigbsq)), each=p)
> x2 = rep(1:p, n)               # 1 2 ... p 1 2 ... p ...
> x3 = runif(p*n)
> linpred = beta[1] + beta[2]*x1 + beta[3]*x2 + beta[4]*x3 + randint
```

```
> expit = exp(linpred)/(1 + exp(linpred))
> y = runif(p*n) < expit          # generate a logical as our outcome
```

We fit the model using the `glmer()` function from the `lme4` package.

```
> library(lme4)

Loading required package: Matrix
Loading required package: Rcpp

> glmmres = glmer(y ~ x1 + x2 + x3 + (1|id), family=binomial(link="logit"))
> summary(glmmres)
Generalized linear mixed model fit by maximum likelihood (Laplace
  Approximation) [glmerMod]
 Family: binomial  ( logit )
Formula: y ~ x1 + x2 + x3 + (1 | id)

     AIC      BIC   logLik deviance df.resid
    5251     5283    -2621     5241     4495

Scaled residuals:
   Min      1Q Median      3Q     Max
-2.019 -0.494 -0.286   0.569   2.846

Random effects:
 Groups Name        Variance Std.Dev.
 id     (Intercept) 3.09     1.76
Number of obs: 4500, groups:  id, 1500

Fixed effects:
            Estimate Std. Error z value Pr(>|z|)
(Intercept)  -1.9668     0.1633  -12.04  < 2e-16 ***
x1            1.5557     0.1319   11.80  < 2e-16 ***
x2            0.4631     0.0501    9.25  < 2e-16 ***
x3           -1.0337     0.1550   -6.67  2.5e-11 ***
---
Signif. codes:  0 '***' 0.001 '**' 0.01 '*' 0.05 '.' 0.1 ' ' 1

Correlation of Fixed Effects:
   (Intr) x1     x2
x1 -0.498
x2 -0.673  0.103
x3 -0.387 -0.073 -0.050
```

10.1.4 Generate correlated binary data

Another way to generate correlated dichotomous outcomes Y_1 and Y_2 is based on the probabilities corresponding to the 2×2 table. Given these cell probabilities, the variable probabilities can be expressed as a function of the marginal probabilities and the desired correlation, using the methods of Lipsitz and colleagues [103]. Here we generate a sample of 1000 values where: $P(Y_1 = 1) = .15$, $P(Y_2 = 1) = .25$, and $\mathrm{Corr}(Y_1, Y_2) = 0.40$.

```
> p1 = .15; p2 = .25; corr = 0.4; n = 10000
> p1p2 = corr*sqrt(p1*(1-p1)*p2*(1-p2)) + p1*p2
> library(Hmisc)
> vals = rMultinom(matrix(c(1-p1-p2+p1p2, p1-p1p2, p2-p1p2, p1p2),
    nrow=1, ncol=4), n)
> y1 = rep(0, n); y2 = rep(0, n)        # put zeroes everywhere
> y1[vals==2 | vals==4] = 1             # and replace them with ones
> y2[vals==3 | vals==4] = 1             # where needed
> rm(vals, p1, p2, p1p2, corr, n)       # cleanup
```

The generated data is close to the desired values.

```
> cor(y1, y2)
[1] 0.429
> table(y1)
y1
   0    1
8515 1485
> table(y2)
y2
   0    1
7542 2458
```

10.1.5   Generate data from a Cox model

To simulate data from a Cox proportional hazards model (7.5.1), we need to model the hazard functions for both time to event and time to censoring. In this example, we use a constant baseline hazard, but this can be modified by specifying other `scale` parameters for the Weibull random variables.

```
> # generate data from Cox model
> n = 10000
> beta1 = 2; beta2 = -1
> lambdaT = .002 # baseline hazard
> lambdaC = .004  # hazard of censoring
> x1 = rnorm(n)    # standard normal
> x2 = rnorm(n)
> # true event time
> T = rweibull(n, shape=1, scale=lambdaT*exp(-beta1*x1-beta2*x2))
> C = rweibull(n, shape=1, scale=lambdaC)   #censoring time
> time = pmin(T,C)  #observed time is min of censored and true
> censored = (time==C)   # set to 1 if event is censored
> # fit Cox model
> library(survival)
> survobj = coxph(Surv(time, (1-censored))~ x1 + x2, method="breslow")
```

These parameters generate data where approximately 40% of the observations are censored. The `coxph()` function expects an observed event indicator. We tabulate the censoring indicator, then display the results as well as the associated 95% confidence intervals.

```
> table(censored)
censored
FALSE  TRUE
 5968  4032
> print(survobj)
Call:
coxph(formula = Surv(time, (1 - censored)) ~ x1 + x2, method = "breslow")


      coef exp(coef) se(coef)     z p
x1  2.02     7.549   0.0224  90.1 0
x2 -1.01     0.363   0.0159 -63.7 0

Likelihood ratio test=11623  on 2 df, p=0  n= 10000, number of events= 5968
> confint(survobj)
    2.5 % 97.5 %
x1  1.98  2.065
x2 -1.05 -0.983
```

The results are similar to the true parameter values.

10.1.6 Sampling from a challenging distribution

When the cumulative density function for a probability distribution can be inverted, it is simple to draw a sample from the distribution using the probability integral transform (3.1.10). However, when the form of the distribution is complex, this approach may be difficult or impossible. The Metropolis–Hastings algorithm [112] is a Markov Chain Monte Carlo (MCMC) method for obtaining samples from a variable with an arbitrary probability density function.

The MCMC algorithm uses a series of draws from a more common distribution, choosing at random which of these proposed draws to accept as draws from the target distribution. The probability of acceptance is calculated so that after the process has converged the accepted draws form a sample from the desired distribution. A further discussion can be found in Section 11.3 of *Probability and Statistics: The Science of Uncertainty* [35] or Section 1.9 of Gelman et al. [50].

Evans and Rosenthal [35] consider a distribution with probability density function:

$$f(y) = c \exp(-y^4)(1 + |y|)^3,$$

where c is a normalizing constant and y is defined on the whole real line.

We find the acceptance probability $\alpha(x, y)$ in terms of two densities, the desired $f(y)$ and $q(x, y)$, a proposal density. For the proposal density, we use the normal with mean equal to the previous value and unit variance, and find that

$$\alpha(x, y) = \min\left\{1, \frac{f(y)q(y, x)}{f(x)q(x, y)}\right\}$$

$$= \min\left\{1, \frac{c \exp\left(-y^4\right)(1 + |y|)^3 (2\pi)^{-1/2} \exp\left(-(y - x)^2/2\right)}{c \exp\left(-x^4\right)(1 + |x|)^3 (2\pi)^{-1/2} \exp\left(-(x - y)^2/2\right)}\right\}$$

$$= \min\left\{1, \frac{\exp\left(-y^4 + x^4\right)(1 + |y|)^3}{(1 + |x|)^3}\right\}.$$

To begin the process, we pick an arbitrary value for X_1. The Metropolis–Hastings algorithm chooses X_{n+1} as follows:

1. Generate y from a normal(X_n, 1).

2. Compute $\alpha(x, y)$ as above.

3. With probability $\alpha(x, y)$, let $X_{n+1} = y$ (use proposal value). Otherwise, with probability $1 - \alpha(x, y)$, let $X_{n+1} = X_n = x$ (repeat previous value).

The code below uses the first 5,000 iterations as a burn-in period, then generates 50,000 samples using this procedure. Only every 10th value from these 50,000 is saved, to reduce autocorrelation. This process is known as "thinning." We begin by writing a function to calculate the acceptance probability.

```
> alphafun = function(x, y) {
     return(exp(-y^4+x^4)*(1+abs(y))^3*
        (1+abs(x))^-3)
  }
```

We generate the samples by using a `for()` loop.

```
> burnin=5000; numvals=5000; thin = 10
> xn = numeric(burnin + numvals*thin)
> xn[1] = rnorm(1)          # starting value
> for (i in 2:(burnin + numvals*thin)) {
     propy = rnorm(1, xn[i-1], 1)
     alpha = min(1, alphafun(xn[i-1], propy))
     xn[i] = sample(c(propy, xn[i-1]), 1, prob=c(alpha, 1-alpha))
  }
> sample = xn[5000 + (1:numvals) * 10]
```

We can compare the true distribution to the empirical PDF estimated from the random draws. To do that, we need the normalizing constant c, which we calculate using R to integrate the distribution over the positive real line.

```
> f = function(x) {
     exp(-x^4)*(1+abs(x))^3
  }
> integral = integrate(f, 0, Inf)
> c = 2 * integral$value; c
[1] 6.81
> pdfeval = function(x) {
     return(1/6.809610784*exp(-x^4)*(1+abs(x))^3)
  }
```

We find that $c = 6.81$.

The results are displayed in Figure 10.1, with the dashed line indicating the true distribution, and the solid line the empirical PDF estimated from the simulated variates. The normalizing constant is used to plot density using the `curve()` function.

Care is always needed when using MCMC methods. This example was particularly well-behaved, in that the proposal distribution is large compared to the distance between the two modes. Section 6.2 of Lavine [94] and Gelman et al. [50] provide accessible discussions of dangers and diagnostics.

```
> curve(pdfeval, from=-2, to=2, lwd=2, lty=2, type="l",
    ylab="probability", xlab="Y")
> lines(density(sample), lwd=2, lty=1)
> legend(-1, .15, legend=c("True", "Simulated"),
    lty=2:1, lwd=2, cex=1.4, bty="n")
```

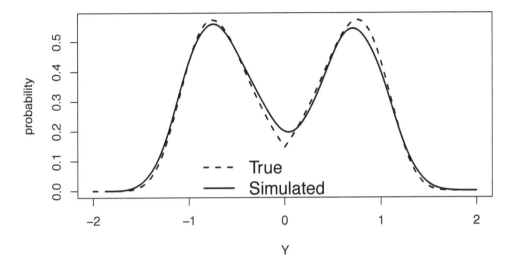

Figure 10.1: Plot of true and simulated distributions

10.2 Simulation applications

10.2.1 Simulation study of Student's t-test

A common rule of thumb is that the sampling distribution of the mean is approximately normal for samples of 30 or larger. Tim Hesterberg has argued (http://home.comcast. net/~timhesterberg/articles/JSM08-n30.pdf) that the $n \geq 30$ rule may need to be retired. He demonstrates that for sample sizes much larger than typically thought necessary, the one-sided error rate of the t-test is not appropriately preserved. We explore this by generating $n = 500$ random exponential variables with mean 1, and carrying out a one-sample Student's t-test. We'll repeat this 1000 times, and examine the coverage of the lower and upper confidence limits.

We first generate the data and perform the test in a `for()` loop. The `foreach` package could be used to speed up the computations if multiple cores were available.

```
> set.seed(42)
> numsim = 1000; n=500
> lower = numeric(numsim); upper = numeric(numsim)
> for (i in 1:numsim) {
    x = rexp(n, rate=1)    # skewed
    testresult = t.test(x, mu=1)
    lower[i] = testresult$conf.int[1] > 1
    upper[i] = testresult$conf.int[2] < 1
  }
```

We use the `tally()` function (5.1.8) to display the proportions of times that the true parameter is not captured.

```
> library(mosaic)
> tally(~ lower, format="percent")


   0    1
97.8  2.2
> tally(~ upper, format="percent")


   0    1
97.3  2.7
```

We observe that the test is rejecting more than it should be on the upper end of the interval (though the overall two-sided alpha level is preserved). An alternative approach would be to generate a matrix of random exponentials, and then use `apply()` to process the computations without a `for` loop (results not shown).

```
> runttest = function(x) {return(CI=confint(t.test(x, mu=1)))}
> xmat = matrix(rexp(numsim*n), nrow=n)
> results = apply(xmat, 2, runttest)
> tally(~ results[2,] > 1)
> tally(~ results[3,] < 1)
```

10.2.2   Diploma (or hat-check) problem

Smith College is a selective women's liberal arts college in Northampton, MA. One tradition acquired since the college was founded in 1871 is to give every graduating student a diploma at random (or more accurately, in a haphazard fashion). At the end of the ceremony, a circle is formed, and students repeatedly pass the diplomas to the person next to them, stepping out once they've received their own diploma. This problem, also known as the *hat-check* problem, is featured in Mosteller [116]. Variants provide great fodder for probability courses.

The analytic solution for the expected number of students who receive their diplomas in the initial disbursement is very straightforward. Let X_i be the event that the ith student receives their diploma. $E[X_i] = 1/n$ for all i, since the diplomas are assumed uniformly distributed. If T is defined as the sum of all of the events X_1 through X_n, $E[T] = n*1/n = 1$ by the rules of expectations. It is sometimes surprising to students that this result does not depend on n. The variance is trickier, since the outcomes are not independent (if one student receives their diploma, the probability that others will increases).

Simulation-based problem solving is increasingly common as a means to complement and enhance analytic solutions [114, 67]. Here we simulate the number of students who receive their diploma and calculate the variance of that quantity. We simulate $n = 650$ students and repeat the experiment 10,000 times. This gives the 650 diplomas a random order, within each simulation. Next, we assign student ID numbers sequentially within the randomly ordered (with respect to diploma number) dataset and mark whether the diploma number matches the student ID. Finally, we count how many times the diploma matches the desired student.

We define a function to carry out one of the simulations. The `students` vector is generated and then permuted using the `sample()` function (see 2.3.3) to represent the diplomas received by the ordered students. The `==` operator (A.4.2) is used to compare each of the elements of the vectors, and the number of matches is returned. The `replicate()` function is used to run this multiple times and save the result.

```
> givedips = function(n) {
    students = 1:n
    diploma = sample(students, n)
    return(sum(students==diploma))
  }
> res = replicate(10000, givedips(650))
```

```
> table(res)
res
   0    1    2    3    4    5    6
3714 3745 1749  612  147   26    7
> library(mosaic)
> favstats(res)
 min Q1 median Q3 max  mean    sd     n missing
   0  0      1  2   6 0.984 0.991 10000       0
```

The expected value and standard deviation of the number of students who receive their diplomas in the random disbursement are both about 1.

10.2.3 Monty Hall problem

The Monty Hall problem illustrates a simple setting where intuition is often misleading. The situation is based on the TV game show *Let's Make a Deal*. First, Monty (the host) puts a prize behind one of three doors. Then the player chooses a door. Next (without moving the prize), Monty opens an unselected door, revealing that the prize is not behind it. The player may then switch to the other nonselected door. Should the player switch?

Many people see that there are now two doors to choose between and feel that since Monty can always open a nonprize door, there's still equal probability for each door. If that were the case, the player might as well keep the original door. This intuition is so attractive that when Marilyn vos Savant asserted that the player should switch (in her *Parade* magazine column), there were reportedly 10,000 letters asserting she was wrong.

A correct intuitive route is to observe that Monty's door is fixed. The probability that the player has the right door is 1/3 before Monty opens the nonprize door, and remains 1/3 after that door is open. This means that the probability the prize is behind one of the other doors is 2/3, both before and after Monty opens the nonprize door. After Monty opens the nonprize door, the player gets a 2/3 chance of winning by switching to the remaining door. If the player wants to win, they should switch doors.

One way to prove to yourself that switching improves your chances of winning is through simulation. In fact, even deciding how to code the problem may be enough to convince yourself to switch.

In the simulation, we assign the prize to a door, then make an initial guess. If the guess was right, Monty can open either door. We'll switch to the other door. Rather than have Monty choose a door, we'll choose one, under the assumption that Monty opened the other one. If our initial guess was wrong, Monty will open the only remaining nonprize door, and when we switch we'll be choosing the prize door.

We write two helper functions. In one, Monty opens a door, choosing at random among the nonchosen doors if the initial choice was correct, or choosing the one nonselected non-prize door if the initial choice was wrong. The other function returns the door chosen by swapping. We use the `sample()` function (2.3.3) to randomly pick one value. We then use these functions on each trial with the `apply()` statement (A.5.2).

```
> numsim = 10000
> doors = 1:3
> opendoor = function(x) {
      # input x is a vector with two values
      # first element is winner, second is choice
      if (x[1]==x[2])    # guess was right
         return(sample(doors[-c(x[1])], 1))
      else return(doors[-c(x[1],x[2])])
  }
```

```
> opendoor(c(1, 1))    # can return 2 or 3
[1] 3
> opendoor(c(1, 2))    # must return 3!
[1] 3
```

Recall that Monty can choose either door 2 or 3 to open when the winning door is initially chosen. When the winning door and initial choice differ (as in the latter example), there is only one door that can be opened.

```
> swapdoor = function(x) { return(doors[-c(x[1], x[2])]) }
> swapdoor(c(1,1))
[1] 2 3
> swapdoor(c(1,2))
[1] 3
```

The `swapdoor()` function works in a similar fashion. Once these parts are in place, the simulation is straightforward.

```
> winner = sample(doors, numsim, replace=TRUE)
> choice = sample(doors, numsim, replace=TRUE)
> open = apply(cbind(winner, choice), 1, opendoor)
> newchoice = apply(cbind(open, choice), 1, swapdoor)
```

```
> cat("Without switching, won ",
      round(sum(winner==choice)/numsim*100, 1),
        " percent of the time.\n", sep="")
Without switching, won 33 percent of the time.
> cat("With switching, won ",
      round(sum(winner==newchoice)/numsim*100, 1),
        " percent of the time.\n", sep="")
With switching, won 67 percent of the time.
```

We note (with some amusement) that Monty didn't actually offer this choice to the players: see `http://tinyurl.com/montynoswitch` for an interview with the details.

10.2.4 Censored survival

A similar problem demonstrates the ways that empirical simulations can complement analytic (closed-form) solutions. Consider an example where a recording device that measures remote activity is placed in a remote location. The time, T, to failure of the remote device has an exponential distribution with mean of 3 years. Since the location is so remote, the device will not be monitored during its first two years of service. As a result, the time to discovery of its failure is $X = \max(T, 2)$. The problem here is to determine the expected value of the observed variable X (e.g., we need to find E[X]).

The analytic solution is fairly straightforward. We need to evaluate:

$$E[X] = \int_0^2 2 * f(u)du + \int_2^\infty u * f(u)du,$$

where $f(u) = 3\exp(-3u)$ for $u > 0$. We can use the calculus functions in the `mosaic` package (Section 3.2.8) to find the answer.

```
> options(digits=6)
> library(mosaic)
> rate = 1/3
> F1 = antiD((lambda*exp(-lambda*t)) ~ t, lambda=rate) # f(T)
> F2 = antiD((t*lambda*exp(-lambda*t)) ~ t, lambda=rate) # E[T]
> 2*(F1(t=2) - F1(t=0)) + (F2(t=Inf) - F2(t=2))
[1] 3.54025
```

It's also straightforward to simulate to confirm the answer.

```
> numsim = 100000
> fail = rexp(numsim, rate=rate)
> # map all values less than 2 to be 2
> fail[fail<2] = 2   # or mean(pmax(2, fail))
> confint(t.test(~ fail))
mean of x    lower    upper    level
   3.53548  3.51926  3.55170  0.95000
```

10.3 Further resources

Rizzo [136] provides a comprehensive introduction to statistical computing in R, while [68] and [67] describe the use of R for simulation studies.

Chapter 11

Special topics

In this chapter, we demonstrate some key programming and statistical techniques that statisticians may encounter in daily practice.

11.1 Processing by group

One of the most common needs in analytic practice is to replicate analyses for subgroups within the data. For example, one may need to stratify a linear regression by gender or repeat a modeling exercise multiple times for each replicate in a simulation experiment. The basic tools for replication in base R include the `by()` function and the `apply()` family of functions (A.5.2). The syntax for these functions can be complicated, however, and various packages exist that can replicate and enhance the functionality provided by `apply()`. One of these is the `dplyr` package developed by Hadley Wickham, demonstrated below; another is the `doBy` package.

11.1.1 Means by group

Examples: 5.7.4 and 2.6.4

The simplest and possibly most common task is to generate simple statistics by a grouping variable. Here we simulate some data to demonstrate. We first use the `tapply()` function.

```
> options(digits=3)
> cat = as.factor(rep(c("blue","red"), each=50))
> y = rnorm(100)
> tapply(y, cat, mean)
   blue     red
-0.0335  0.2227
```

or

```
> xtabs(~ ave(y, cat, FUN=mean))
ave(y, cat, FUN = mean)
-0.0334599326462414    0.222658846376363
                50                    50
```

or

```
> library(mosaic)
> mean(y ~ cat)
   blue      red
-0.0335   0.2227
```

or

```
> library(dplyr)
> ds = data.frame(y, cat)
> groups = group_by(ds, cat)
> summarise(groups, mean=mean(y))
Source: local data frame [2 x 2]

   cat     mean
1 blue -0.0335
2  red  0.2227
```

The `tapply()` function applies the function given as the third argument (in this case `mean()`) to the vector in the first argument (`y`) stratified by every unique set of values of the list of factors in the second argument (`x`). It returns a vector of that length with the results of the function. Similar functionality is available using the `by()` or `ave()` functions (see `example(ave)`), the latter of which returns a vector of the same length as `x` with each element equal to the mean of the subset of observations with the factor level specified by `y`. Many functions (e.g., `mean()`, `median()` and `favstats()`) within the `mosaic` package support a lattice-style modeling language for summary statistics. Finally, the `group_by()` and `summarise()` functions in the `dplyr` package provide a powerful mechanism for grouped operations.

11.1.2 Linear models stratified by each value of a grouping variable

We'll use the HELP data to assess the relationship between age and drinking, by gender. We begin by showing a way to do this from scratch, i.e., without convenience functions. It's often a useful programming exercise to code such routines, rather than relying on existing functions or packages.

```
> ds = read.csv("http://www.amherst.edu/~nhorton/r2/datasets/help.csv")
> uniquevals = unique(ds$female)
> numunique = length(uniquevals)
> formula = as.formula(i1 ~ age)
> p = length(coef(lm(formula, data=ds)))
> params = matrix(rep(0, numunique*p), nrow=p, ncol=numunique)
> for (i in 1:length(uniquevals)) {
    cat("grouping:", i, "\n")
      params[,i] = coef(lm(formula, data=subset(ds,
        female==uniquevals[i])))
  }
grouping: 1
grouping: 2
```

```
> rownames(params) = c("Intercept", "Age")
> colnames(params) = ifelse(uniquevals==0, "male", "female")
> params
          male female
Intercept -3.693  5.829
Age        0.635  0.251
```

In the above code, separate regressions are fit for each value of the grouping variable z through use of a `for` loop. This requires the creation of a matrix of results `params` to be set up in advance, of the appropriate dimension (number of rows equal to the number of parameters ($p = k + 1$) for the model, and number of columns equal to the number of levels for the grouping variable z). Within the loop, the `lm()` function is called, and the coefficients from each fit are saved in the appropriate column of the `params` matrix.

A simpler and more elegant approach is to use the `dlply()` and `ldply()` functions from Hadley Wickham's `plyr` package.

```
> library(plyr)
> models = dlply(ds, "female", function(df) {
    lm(i1 ~ age, data=df)
  })
> ldply(models, coef)
  female (Intercept)   age
1      0        -3.69 0.635
2      1         5.83 0.251
```

The `dlply()` function splits a data frame, applies a function to each of the parts, and returns the results in a list. The `ldply()` function reverses this process: it splits a list, applies a function to each element, and returns a data frame. Note that we define the function within the call to `dlply()`, without giving it a name. This is often a useful technique in `apply()`-like functions (A.5.2).

11.2 Simulation-based power calculations

In some settings, analytic power calculations (5.5) may not be readily available. A straightforward alternative is to estimate power empirically, simulating data from the proposed design under given assumptions regarding the alternative.

We consider a study of children clustered within families. Each family has three children; in some families all three children have an exposure of interest, while in others just one child is exposed. In the simulation, we assume that the outcome is multivariate normal with higher mean for those with the exposure, and 0 for those without. A compound symmetry correlation is imposed, with equal variances for each child. We assess the power to detect an exposure effect where the intended analysis uses a random intercept model (7.4.2) to account for the clustering within families. With this simple covariance structure, it is trivial to generate correlated errors directly. We specify the correlation matrix directly and simulate the multivariate normal.

```
> library(MASS)
> library(nlme)
> # initialize parameters and building blocks
> effect = 0.35   # effect size
```

```
> corr = 0.4          # intrafamilial correlation
> numsim = 1000
> n1fam = 50          # families with 3 exposed
> n2fam = 50          # families with 1 exposed and 2 unexposed
> # 3x3 compound symmetry matrix
> vmat = matrix(c
    ( 1,    corr, corr,
     corr, 1   , corr,
     corr, corr, 1    ), nrow=3, ncol=3)
> # 1 1 1 ... 1 0 0 0 ... 0
> x = c(rep(1, n1fam), rep(1, n1fam), rep(1, n1fam),
         rep(1, n2fam), rep(0, n2fam), rep(0, n2fam))
> # 1 2 ... n1fam 1 2 ... n1fam ...
> id = c(1:n1fam, 1:n1fam, 1:n1fam,
    (n1fam+1:n2fam), (n1fam+1:n2fam), (n1fam+1:n2fam))
> power = rep(0, numsim) # initialize vector for results
```

The concatenate function (c()) is used to glue together the appropriate elements of the design matrices and underlying correlation structure.

```
> for (i in 1:numsim) {
    # all three exposed
    grp1 = mvrnorm(n1fam, c(effect, effect, effect), vmat)

    # only first exposed
    grp2 = mvrnorm(n2fam, c(effect, 0,       0),       vmat)

    # concatenate the output vector
    y = c(grp1[,1], grp1[,2], grp1[,3],
          grp2[,1], grp2[,2], grp2[,3])

    group = groupedData(y ~ x | id)   # specify dependence structure
    res = lme(group, random = ~ 1)    # fit random intercept model
    pval = summary(res)$tTable[2,5]   # grab results for main parameter
    power[i] = pval<=0.05             # is it statistically significant?
}
```

The proportion of rejections is the empirical estimate of power. This yields the following power estimate (and confidence interval due to simulation).

```
> cat("\nEmpirical power for effect size of ", effect,
    " is ", round(sum(power)/numsim,3), ".\n", sep="")

Empirical power for effect size of 0.35 is 0.841.
> cat("95% confidence interval is",
    round(prop.test(sum(power), numsim)$conf.int, 3), "\n")
95% confidence interval is 0.817 0.863
```

11.3 Reproducible analysis and output

The key idea of reproducible analysis is that data analysis code, results, and interpretation should all be located together. This stems from the concept of "literate programming" (in the sense of Knuth [86]) and facilitates transparent and repeatable analysis [95, 52]. Reproducible analysis systems, which are becoming more widely adopted [13], help to provide a clear audit trail and automate report creation. Ultimately, the goal is to avoid post-analysis cut-and-paste processing, which has a high probability of introducing errors.

There are various implementations of reproducible analysis in R [95, 199], several of which made the production of this book possible. Each of these systems functions by allowing the analyst to combine code and text into a single file. This file is processed to extract the code, run it through the statistical systems in batch mode, collect the results, then integrate the text, code, output, and graphical displays into the final document. The systems available in R are extensive and are an active area of development.

The most powerful and flexible system is the `knitr` package (due to Yihui Xie [199]). The package can be used by writing a file in the LaTeX document markup language, but another useful option is to write it in the far simpler Markdown format. Markdown files can be converted to a variety of common display and editing formats, such as PDF and Microsoft Word, using Pandoc (`http://johnmcfarlane.net/pandoc`, a "Swiss Army knife" of file conversion).

The `knitr` package is well-integrated with RStudio, and both LaTeX/PDF and Markdown/Pandoc conversions to several formats are provided via single-click mechanisms. More details can be found in [199] and [47] as well as the CRAN reproducible analysis task view (see also `http://yihui.name/knitr`).

As an example of how these systems work, we demonstrate a document written in the Markdown format using data from the built-in `cars` data frame. Within RStudio, a new template R Markdown file can be generated by selecting `R Markdown` from the `New File` option on the `File` menu. This generates the dialog box displayed in Figure 11.1. The default output format is HTML, but other options are available.

Figure 11.2 displays this default Markdown input file. The file is given a title (`Sample R Markdown example`) with output format set by default to HTML. Simple markup (such as bolding) is added through use of the `**` characters before and after the word `Help`. Blocks of code are begun using the ```{r} command and closed with a ``` command (three back quotes). In this example, the correlation between two variables is calculated and a scatterplot is generated.

The formatted output can be generated and displayed by clicking the `Knit HTML` button in RStudio, or by using the commands in the following code block, which can also be used when running R without the benefit of RStudio.

```
> library(markdown); library(knitr)
> knit("filename.Rmd")   # creates filename.md
> markdownToHTML("filename.md", "filename.html")
> browseURL("filename.html")
```

The `knit()` function extracts the R commands from a specially formatted R Markdown input file (`filename.Rmd`), evaluates them, and integrates the resulting output, including text and graphics, into an intermediate file (`filename.md`). This file is then processed (using `markdownToHTML()`) to create a final display file in HTML format. A screenshot of the results of performing these steps on the `.Rmd` file displayed in Figure 11.2 is displayed in Figure 11.3.

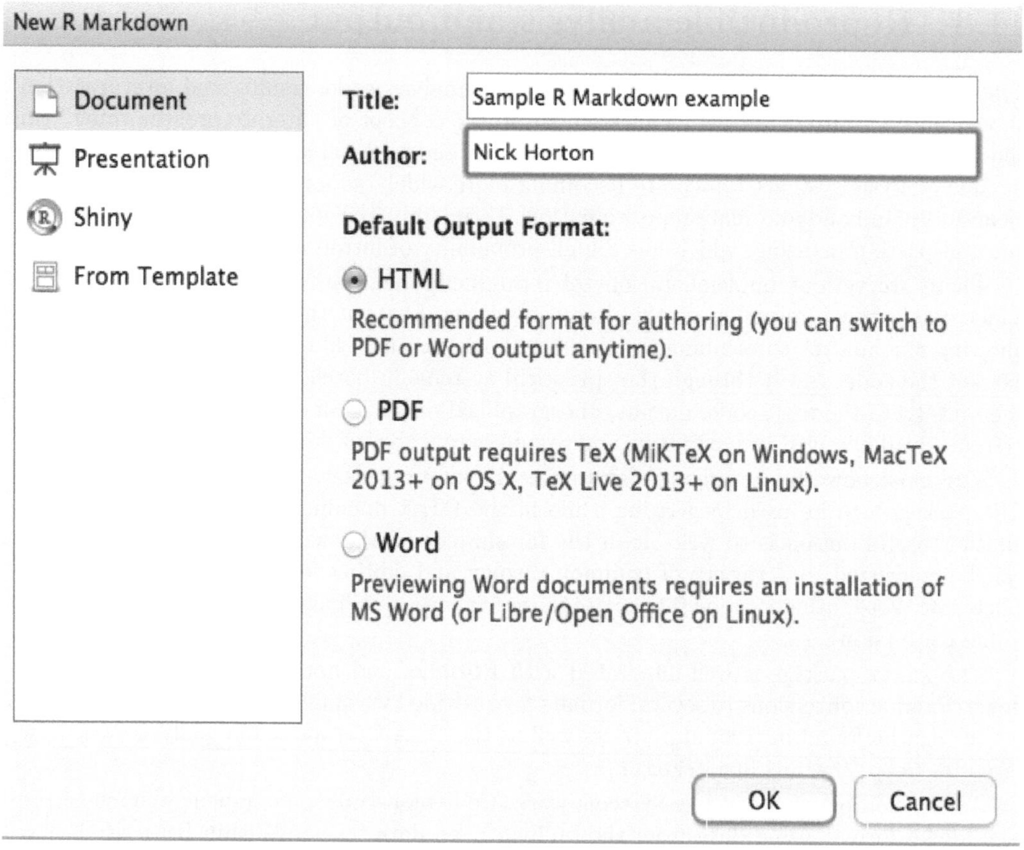

New R Markdown

| Document |
| Presentation |
| Shiny |
| From Template |

Title: Sample R Markdown example

Author: Nick Horton

Default Output Format:

◉ **HTML**

Recommended format for authoring (you can switch to PDF or Word output anytime).

○ **PDF**

PDF output requires TeX (MiKTeX on Windows, MacTeX 2013+ on OS X, TeX Live 2013+ on Linux).

○ **Word**

Previewing Word documents requires an installation of MS Word (or Libre/Open Office on Linux).

[OK]   [Cancel]

Figure 11.1: Generating a new R Markdown file in RStudio

The `knit()` function operates, by default, on the convention that input files ending with `.Rmd` generate a `.md` (Markdown) file, and files ending with `.Rnw` generate a `.tex` (LaTeX) file.

Alternatively, a PDF or Microsoft Word file can be generated in RStudio by selecting `New` from the `R Markdown` menu, then clicking on the PDF or Word options. RStudio also supports the creation of R Presentations using a variant of the R Markdown language. Instructions and an example can be found by opening a new `R presentations` document in RStudio.

A LaTeX file can be generated using the following commands, where `filename.Rnw` is a LaTeX file with specific codes indicating the presence of R statements.

```
> library(knitr)
> knit("filename.Rnw")
```

The resulting `filename.tex` file could then be compiled with `pdflatex` in the operating system, resulting in a PDF file. This is done automatically using the `Compile to PDF` button in RStudio.

It's often useful to evaluate the code separately. The `Stangle()` function creates a file containing the code chunks and omitting the text. The resulting file could be run as a script using `source()`, and would generate just the results seen in the woven document.

```
---
title: "Sample R Markdown example"
author: "Nick Horton"
date: "October 4, 2014"
output: html_document
---
```

This is an R Markdown document. Markdown is a simple formatting syntax for authoring HTML, PDF, and MS Word documents. For more details on using R Markdown see <http://rmarkdown.rstudio.com>.

When you click the **Knit** button a document will be generated that includes both content as well as the output of any embedded R code chunks within the document. You can embed an R code chunk like this:

```
'''{r}
summary(cars)
'''
```

You can also embed plots, for example:

```
'''{r, echo=FALSE}
plot(cars)
'''
```

Note that the `echo = FALSE` parameter was added to the code chunk to prevent printing of the R code that generated the plot.

Figure 11.2: Sample Markdown input file

The `spin()` function in the `knitr` package takes a formatted R script and produces an R Markdown document. This can be helpful for those moving from the use of scripts to more structured Markdown files.

11.4 Advanced statistical methods

In this section, we discuss implementations of modern statistical methods and techniques, including Bayesian methods, propensity score analysis, missing data methods, and estimation of finite mixture models.

11.4.1 Bayesian methods

Bayesian methods are increasingly commonly utilized, and implementations of many models are available in R.

We focus here on Markov Chain Monte Carlo (MCMC) methods for model fitting, which are quite general and much more flexible than closed form solutions. Diagnosis of convergence is a critical part of any MCMC model fitting (see Gelman et. al., [50] for an accessible introduction). Support for model assessment is provided, for example, in the

This is an R Markdown document. Markdown is a simple formatting syntax for authoring HTML, PDF, and MS Word documents. For more details on using R Markdown see http://rmarkdown.rstudio.com.

When you click the **Knit** button a document will be generated that includes both content as well as the output of any embedded R code chunks within the document. You can embed an R code chunk like this:

```
summary(cars)
```

```
##      speed           dist
##  Min.   : 4.0   Min.   :  2
##  1st Qu.:12.0   1st Qu.: 26
##  Median :15.0   Median : 36
##  Mean   :15.4   Mean   : 43
##  3rd Qu.:19.0   3rd Qu.: 56
##  Max.   :25.0   Max.   :120
```

You can also embed plots, for example:

Figure 11.3: Formatted output from R Markdown example

coda (Convergence Diagnosis and Output Analysis) package written by Kate Cowles and others.

```
> library(MCMCpack)
> # linear regression
> mod1 = MCMCregress(formula, burnin=1000, mcmc=10000, data=ds)
> # logistic regression
> mod2 = MCMClogit(formula, burnin=1000, mcmc=10000, data=ds)
> # Poisson regression
> mod3 = MCMCpoisson(formula, burnin=1000, mcmc=10000, data=ds)
```

The CRAN task view on Bayesian inference provides an overview of the packages that incorporate some aspect of Bayesian methodologies. Table 11.1 displays modeling functions available within the MCMCpack package (including the three listed above). By default, the prior mean and precision are set to 0, equivalent to an improper uniform distribution.

More general MCMC models can also be fit in R, typically in packages that call stand-alone MCMC software such as OpenBUGS, JAGS, or WinBUGS. These packages include BRugs, R2WinBUGS, rjags, R2jags, and runjags.

Table 11.1: Bayesian modeling functions available within the `MCMCpack` package

MCMCbinaryChange()	MCMC for a Binary Multiple Changepoint Model
MCMCdynamicEI()	MCMC for Quinn's Dynamic Ecological Inference Model
MCMCdynamicIRT1d()	MCMC for Dynamic One-Dimensional Item Response Theory Model
MCMCfactanal()	MCMC for Normal Theory Factor Analysis Model
MCMChierEI()	MCMC for Wakefield's Hierarchical Ecological Inference Model
MCMCirt1d()	MCMC for One-Dimensional Item Response Theory Model
MCMCirtHier1d()	MCMC for Hierarchical One-Dimensional Item Response Theory Model, Covariates Predicting Latent Ideal Point (Ability)
MCMCirtKd()	MCMC for K-Dimensional Item Response Theory Model
MCMCirtKdHet()	MCMC for Heteroskedastic K-Dimensional Item Response Theory Model
MCMCirtKdRob()	MCMC for Robust K-Dimensional Item Response Theory Model
MCMClogit()	MCMC for Logistic Regression
MCMCmetrop1R()	Metropolis Sampling from User-Written R function
MCMCmixfactanal()	MCMC for Mixed Data Factor Analysis Model
MCMCmnl()	MCMC for Multinomial Logistic Regression
MCMCoprobit()	MCMC for Ordered Probit Regression
MCMCordfactanal()	MCMC for Ordinal Data Factor Analysis Model
MCMCpoisson()	MCMC for Poisson Regression
MCMCpoissonChange()	MCMC for a Poisson Regression Changepoint Model
MCMCprobit()	MCMC for Probit Regression
MCMCquantreg()	Bayesian Quantile Regression Using Gibbs Sampling
MCMCregress()	MCMC for Gaussian Linear Regression
MCMCSVDreg()	MCMC for SVD Regression
MCMCtobit()	MCMC for Gaussian Linear Regression with a Censored Dependent Variable

11.4.1.1 Logistic regression via MCMC

One use for Bayesian logistic regression might be in the case of complete or quasi-complete separation. Loosely, this occurs when all the subjects in some level of the exposure variables have the same outcome status. Here, we simulate such data and demonstrate how to use Bayesian MCMC methods to fit the model. The simulated data have 100 trials in each of two levels of a predictor, with 0 and 5 events in the two levels. Note that the classical estimated odds ratio is infinity, or undefined, and that different software implementations behave unpredictably in this instance.

```
> events.0=0    # for X = 0
> events.1=5    # for X = 1
> x = as.factor(c(rep(0,100), rep(1,100)))
> y = c(rep(0,100-events.0), rep(1,events.0),
    rep(0, 100-events.1), rep(1, events.1))
>
> library(MCMCpack)
> logmcmc = MCMClogit(y ~ x, burnin=100, mcmc=2000, b0=0, B0=.04)
```

```
> summary(logmcmc)

Iterations = 101:2100
Thinning interval = 1
Number of chains = 1
Sample size per chain = 2000

1. Empirical mean and standard deviation for each variable,
   plus standard error of the mean:

             Mean   SD Naive SE Time-series SE
(Intercept) -6.50 1.70   0.0381         0.0931
x1           3.42 1.77   0.0395         0.1009

2. Quantiles for each variable:

             2.5%   25%   50%   75% 97.5%
(Intercept) -10.606 -7.48 -6.26 -5.30 -3.97
x1            0.708  2.15  3.15  4.52  7.50
```

Under the default normal prior, the mean of the prior is set with `b0`; `B0` is the prior precision. The `burnin` and `mcmc` options define the number of iterations discarded before inference iterations are captured, and the number of iterations for inference, respectively.

11.4.1.2   Poisson regression

In addition to problematic examples such as the quasi-complete separation seen above, it may be desirable to consider Bayesian techniques in more conventional settings. Here, we demonstrate a Poisson regression using the HELP dataset. We use the `MCMCpoisson()` function.

```
> library(MCMCpack)
> posterior = with(ds,
    MCMCpoisson(i1 ~ female + as.factor(substance) + age,
        burnin=100, mcmc=2000))
```

```
> summary(posterior)

Iterations = 101:2100
Thinning interval = 1
Number of chains = 1
Sample size per chain = 2000

1. Empirical mean and standard deviation for each variable,
   plus standard error of the mean:

                             Mean      SD Naive SE Time-series SE
(Intercept)                2.8888 0.06255 1.40e-03       0.005864
female                    -0.1728 0.03109 6.95e-04       0.002888
as.factor(substance)cocaine -0.8140 0.02817 6.30e-04     0.002329
as.factor(substance)heroin -1.1166 0.03477 7.78e-04      0.003252
age                        0.0135 0.00155 3.47e-05       0.000147

2. Quantiles for each variable:

                             2.5%     25%     50%     75%    97.5%
(Intercept)                2.7703  2.8454  2.8871  2.9349  3.0095
female                    -0.2355 -0.1946 -0.1734 -0.1508 -0.1105
as.factor(substance)cocaine -0.8676 -0.8325 -0.8134 -0.7935 -0.7615
as.factor(substance)heroin -1.1856 -1.1391 -1.1153 -1.0922 -1.0485
age                        0.0105  0.0124  0.0135  0.0146  0.0163
```

Default plots are available for `MCMC` objects returned by `MCMCpack`. These can be displayed using the command `plot(posterior)`.

11.4.2 Propensity scores

Propensity scores can be used to attempt causal inference in an observational setting where there are potential confounding factors [138, 139]. Here we consider comparisons of the physical component scores (PCS) for homeless vs. nonhomeless subjects in the HELP study. Does homelessness make people less physically competent?

First, we examine the observed difference in PCS between homeless and housed.

```
> lm1 = lm(pcs ~ homeless, data=ds)
> summary(lm1)

Call:
lm(formula = pcs ~ homeless, data = ds)

Residuals:
   Min     1Q Median     3Q    Max
-34.93  -7.90   0.64   8.39  25.81

Coefficients:
            Estimate Std. Error t value Pr(>|t|)
(Intercept)   49.001      0.688   71.22  <2e-16 ***
```

```
homeless      -2.064      1.013    -2.04    0.042 *
---
Signif. codes:  0 '***' 0.001 '**' 0.01 '*' 0.05 '.' 0.1 ' ' 1

Residual standard error: 10.7 on 451 degrees of freedom
Multiple R-squared:  0.00912,Adjusted R-squared:  0.00693
F-statistic: 4.15 on 1 and 451 DF,  p-value: 0.0422
```

We see statistically significant lower mean PCS for the homeless ($p = 0.042$). However, subjects were not randomized to homelessness. Homelessness may be a result of confounding factors that are associated with homelessness and cause reduced physical competence. If we want to make causal inference about the effects of homelessness, we need to adjust for these confounders.

11.4.2.1 Regression adjustment

One approach to this problem involves controlling for possible confounders (in this case, age, gender, number of drinks, and MCS score) in a multiple regression model (6.1.1).

```
> lm2 = lm(pcs ~ homeless + age + female + i1 + mcs, data=ds)
> summary(lm2)

Call:
lm(formula = pcs ~ homeless + age + female + i1 + mcs, data = ds)

Residuals:
   Min    1Q Median    3Q    Max
-35.77  -6.67   0.41  7.67  26.59

Coefficients:
            Estimate Std. Error t value Pr(>|t|)
(Intercept) 58.2122     2.5667   22.68  < 2e-16 ***
homeless    -1.1471     0.9979   -1.15  0.25099
age         -0.2659     0.0641   -4.15    4e-05 ***
female      -3.9552     1.1514   -3.44  0.00065 ***
i1          -0.0808     0.0254   -3.18  0.00156 **
mcs          0.0703     0.0381    1.85  0.06540 .
---
Signif. codes:  0 '***' 0.001 '**' 0.01 '*' 0.05 '.' 0.1 ' ' 1

Residual standard error: 10.2 on 447 degrees of freedom
Multiple R-squared:  0.112,Adjusted R-squared:  0.102
F-statistic: 11.2 on 5 and 447 DF,  p-value: 3.21e-10
```

Controlling for the other predictors has caused the parameter estimate to attenuate to the point that it is no longer statistically significant ($p = 0.25$). While controlling for other confounders may be effective in this problem, other situations may be more vexing, particularly if the dataset is small and the number of measured confounders is large. In such settings, the propensity score (the probability of being homeless, conditional on other factors), can be used. Typical applications include regression adjustment for the propensity to exposure, matching on the propensity to exposure, and stratification into similar levels of propensity. Propensity scores also allow easy investigation of the overlap in covariate space, a requirement for effective multiple regression adjustment that is often ignored. Here, we demonstrate estimating the propensities, using them in a regression adjustment, and matching. Assessment of overlap resembles the assessment of the linear discriminant analysis in 7.10.18. For an example of stratification, see http://tinyurl.com/sasrblog-propensity.

11.4.2.2 Estimating the propensity score

A typical way to estimate the propensity score is to model the exposure as a function of covariates in a logistic regression model (7.1.1). We use a `formula` object (see 6.1.1) to specify the model.

```
> form = formula(homeless ~ age + female + i1 + mcs)
> glm1 = glm(form, family=binomial, data=ds)
> propensity = glm1$fitted
```

The `glm1` object has a `fitted` element that contains the predicted probability from the logistic regression. For ease of use, we extract it to a new object.

11.4.2.3 Regression adjustment for propensity

The simplest use of the propensity score is to include it as a continuous covariate in a regression model.

```
> lm3 = lm(pcs ~ homeless + propensity, data=ds)
> summary(lm3)

Call:
lm(formula = pcs ~ homeless + propensity, data = ds)

Residuals:
   Min     1Q Median     3Q    Max
-34.03  -7.62   0.93   8.24  25.65

Coefficients:
            Estimate Std. Error t value Pr(>|t|)
(Intercept)    54.54       1.83   29.88   <2e-16 ***
homeless       -1.18       1.04   -1.14   0.2569
propensity    -12.89       3.94   -3.27   0.0012 **
---
Signif. codes:  0 '***' 0.001 '**' 0.01 '*' 0.05 '.' 0.1 ' ' 1

Residual standard error: 10.6 on 450 degrees of freedom
Multiple R-squared:  0.0321,Adjusted R-squared:  0.0278
F-statistic: 7.47 on 2 and 450 DF,  p-value: 0.000645
```

As with the multiple regression model, controlling for the propensity also leads to an attenuated estimate of the homeless coefficient.

11.4.2.4 Matching on propensity score

Another approach matches exposed and unexposed (homeless and nonhomeless) subjects with similar propensity scores. This typically generates a sample that is approximately balanced on the terms included in the propensity model. Since a confounded effect requires a disequilibrium of the confounders between the groups, this can be an effective treatment. Matching and comparison are straightforward to do using the `Matching` package.

```
> library(Matching)
> rr = with(ds, Match(Y=pcs, Tr=homeless, X=propensity, M=1))
> summary(rr)


Estimate...  -0.80207
AI SE......  1.4448
T-stat.....  -0.55516
p.val......  0.57878


Original number of observations.............. 453
Original number of treated obs.............. 209
Matched number of observations.............. 209
Matched number of observations  (unweighted). 252
```

We see that the causal estimate of -0.80 in the matched comparison is not significantly different from zero ($p = 0.58$), which is similar to the results from the other approaches that accounted for the possible confounders.

11.4.2.5   Assessing balance after matching

It would be wise to make a further investigation of whether the matching "worked," in the sense of making the groups more similar with respect to the potential confounders.

For example, here are the means and standard deviations among the whole sample (including just two covariates for space reasons).

Note that while balance was improved for both covariates, there remains some difference between the groups. The `MatchBalance()` function can be used to describe the distribution of the predictors (by homeless status) before and after matching (to save space, only the results for `age` are displayed).

```
> longout = capture.output(MatchBalance(form, match.out=rr,
    nboots=10, data=ds))
> write(longout[1:20], file = "")


***** (V1) age *****
                        Before Matching      After Matching
mean treatment........      36.368             36.368
mean control..........      35.041             36.423
std mean diff.........      16.069             -0.65642


mean raw eQQ diff.....      1.5981             0.94841
med  raw eQQ diff.....      1                  1
max  raw eQQ diff.....      7                  10


mean eCDF diff........      0.037112           0.022581
med  eCDF diff........      0.026365           0.019841
max  eCDF diff........      0.10477            0.083333
```

```
var ratio (Tr/Co).....      1.329         1.2671
T-test p-value........    0.070785      0.93902
KS Bootstrap p-value..        0.2           0.2
KS Naive p-value......    0.16881       0.34573
KS Statistic.........     0.10477       0.083333
```

The `capture.output()` function is used to send the voluminous output to a character string, so that only a subset can be displayed. After matching, the age variables had distributions that were considerably closer to each other.

The `Match()` function can also be used to generate a dataset containing only the matched observations (see the `index.treated` and `index.control` components of the `Match` object).

11.4.3 Bootstrapping

Bootstrapping is a powerful and elegant approach to estimating the sampling distribution of statistics. It can be implemented even in many situations where asymptotic results are difficult to find or otherwise unsatisfactory [33]. Bootstrapping proceeds using three steps: first, resample the dataset (with replacement) many times (typically on the order of 10,000); then calculate the desired statistic from each resampled dataset; finally, use the distribution of the resampled statistics to estimate the standard error of the statistic (normal approximation method) or construct a confidence interval using quantiles of that distribution (percentile method). There are several ways to bootstrap in R.

As an example, we consider estimating the standard error and 95% confidence interval for the coefficient of variation (CV), defined as σ/μ, for a random variable X. We'll generate normal data with a mean and variance of 1.

```
> x = rnorm(1000, mean=1)
```

The user must provide code to calculate the statistic of interest as a function.

```
> covfun = function(x) {  # multiply CV by 100
    return(100*sd(x)/mean(x))
 }
```

The `replicate()` function is the base R tool for repeating function calls. Here, we nest within it a call to our `covfun()` function and a call to sample the data with replacement using the `sample()` function.

```
> options(digits=4)
> res2 = replicate(2000, covfun(sample(x, replace=TRUE)))
> quantile(res2, c(.025, .975))
  2.5%  97.5%
 98.85 116.07
```

The `do()` function from the `mosaic` package provides an alternative syntax, while its `resample()` function is a convenience function that provides appropriate defaults to `sample`.

```
> options(digits=4)
> covfun(x)
[1] 106.5
> library(mosaic)
> res = do(2000) * covfun(resample(x))
> quantile(res$result, c(.025, .975))
  2.5%  97.5%
 98.31 115.56
```

The percentile interval is simple to calculate from the observed bootstrapped statistics. If the distribution of the bootstrap samples is approximately normally distributed, a t interval could be created by calculating the standard deviation of the bootstrap samples and finding the appropriate multiplier for the confidence interval (more information can be found in the mosaic package resampling vignette). Plotting the bootstrap sample estimates is helpful to determine the form of the bootstrap distribution [65]. The do() function provides a natural syntax for repetition (see 2.3.3 and replicate()). The boot package also provides a rich set of routines for bootstrapping, including support for bias-corrected and accelerated intervals.

11.4.4   Missing data

11.4.4.1   Account for missing values

Missing values are ubiquitous in most real-world investigations. R includes support for missing value codes, though there are important aspects that need to be kept in mind by an analyst, particularly when deriving new variables or fitting models.

Missing values are denoted by NA, a logical constant of length 1 that has no numeric equivalent. The missing value code is distinct from the character string value "NA". The default behavior for most R functions is to return NA if any of the input vectors have any missing values.

```
> x = c(1, 2, NA)
> mean(x)
[1] NA
```

```
> mean(x, na.rm=TRUE)
[1] 1.5
```

```
> sum(na.omit(x))
[1] 3
```

```
> sum(!is.na(x))
[1] 2
```

The na.rm option is used within the mean() function to override the default behavior, omit missing values, and calculate the result on the complete cases. Many other functions allow the specification of an na.action option (e.g., for the lm() function). Common na.action functions include na.exclude(), na.omit(), and na.fail() (see also na.action() and `options("na.action"))`. The `!` (not) Boolean operator allows counting of the number of observed values (since `is.na()` returns a logical set to TRUE if an observation is missing).

The `na.omit()` function returns the dataframe with missing values omitted (if a value is missing for a given row, all observations are removed, aka listwise deletion, see also `naresid()`).

The `scan()` and `read.table()` functions have the default argument `na.strings="NA"`. This can be used to recode on input for situations where a numeric missing value code has been used. The `table()` function provides the `exclude=NULL` option to include a category for missing values. R has other kinds of "missing" values, corresponding to floating-point standards (see also the `is.infinite()` and `is.nan()` functions).

```
> # remap values of x with missing value code of 999 to missing
> w = c(1, 2, 999)
> w[w==999] = NA
> w
[1]  1  2 NA
```

or

```
> w = c(1,2,999)
>
> is.na(w) = w==999 # set 999's to missing
> w
[1]  1  2 NA
```

Arbitrary numeric missing values (999 in this example) can be mapped to R missing value codes using indexing and assignment. Here, all values of x that are 999 are replaced by the missing value code of NA. The `na.pattern()` function in the `Hmisc` package can be used to determine the different patterns of missing values in a dataset.

11.4.4.2 Account for missing data using multiple imputation

Here, we demonstrate some of the capabilities for fitting incomplete data regression models using multiple imputation [142, 149, 70] implemented with chained equation models [177, 131, 176].

In this example, we replicate an analysis from 7.10.1 in a version of the HELP dataset that includes missing values for several of the predictors. While not part of the regression model of interest, the `mcs` and `pcs` variables are included in the imputation models, which may make the missing-at-random assumption more plausible [27].

We begin by reading in the data, then using the `na.pattern()` function from the `Hmisc` package to characterize the patterns of missing values.

```
> ds =
    read.csv("http://www.amherst.edu/~nhorton/r2/datasets/helpmiss.csv")
> smallds = with(ds, data.frame(homeless, female, i1, sexrisk, indtot,
    mcs, pcs))
```

```
> summary(smallds)
   homeless          female             i1            sexrisk
 Min.   :0.000   Min.   :0.000   Min.   :  0.0   Min.   : 0.00
```

```
1st Qu.:0.000    1st Qu.:0.000    1st Qu.:   3.0    1st Qu.:  3.00
Median :0.000    Median :0.000    Median :  13.0    Median :  4.00
Mean   :0.466    Mean   :0.236    Mean   :  18.3    Mean   :  4.63
3rd Qu.:1.000    3rd Qu.:0.000    3rd Qu.:  26.0    3rd Qu.:  6.00
Max.   :1.000    Max.   :1.000    Max.   : 142.0    Max.   : 14.00
                                                    NA's   :1
        indtot           mcs              pcs
Min.   : 4.0     Min.   : 6.76    Min.   :14.1
1st Qu.:32.0     1st Qu.:21.66    1st Qu.:40.4
Median :37.5     Median :28.56    Median :48.9
Mean   :35.7     Mean   :31.55    Mean   :48.1
3rd Qu.:41.0     3rd Qu.:40.64    3rd Qu.:57.0
Max.   :45.0     Max.   :62.18    Max.   :74.8
NA's   :14       NA's   :2        NA's   :2
> library(Hmisc)
> na.pattern(smallds)
pattern
0000000 0000011 0000100 0001100
    454       2      13       1
```

There are 14 subjects missing `indtot`, 2 missing `mcs` as well as `pcs`, and 1 missing `sexrisk`. In terms of patterns of missingness, there are 454 observations with complete data, 2 missing both `mcs` and `pcs`, 13 missing `indtot` alone, and 1 missing `sexrisk` and `indtot`. Fitting a logistic regression model (7.1.1) using the available data ($n = 456$) yields the following results.

```
> glm(homeless ~ female + i1 + sexrisk + indtot, binomial,
      data=smallds)

Call:  glm(formula = homeless ~ female + i1 + sexrisk + indtot,
    family = binomial, data = smallds)

Coefficients:
(Intercept)         female              i1       sexrisk           indtot
    -2.5278        -0.2401          0.0232        0.0562           0.0493

Degrees of Freedom: 455 Total (i.e. Null);   451 Residual
  (14 observations deleted due to missingness)
Null Deviance:      630
Residual Deviance: 586   AIC: 596
```

Next, the `mice()` function within the `mice` package is used to impute missing values for `sexrisk`, `indtot`, `mcs`, and `pcs`. These results are combined using `glm.mids()`, and results are pooled and reported. Note that by default, all variables within the `smallds` data frame are included in each of the chained equations (so that `mcs` and `pcs` are used as predictors in each of the imputation models).

```
> library(mice)
> imp = mice(smallds, m=20, maxit=25, seed=42, print=FALSE)
```

```
> summary(pool(glm.mids(homeless ~ female + i1 + sexrisk +
      indtot, family=binomial, data=imp)))
                 est       se       t     df  Pr(>|t|)      lo 95     hi 95
(Intercept) -2.53050 0.593809 -4.261 450.7 2.474e-05 -3.69748 -1.36352
female      -0.24536 0.243884 -1.006 462.0 3.149e-01 -0.72462  0.23390
i1           0.02313 0.005615  4.119 462.4 4.515e-05  0.01209  0.03416
sexrisk      0.05975 0.035810  1.669 461.4 9.587e-02 -0.01062  0.13012
indtot       0.04888 0.015796  3.095 447.6 2.094e-03  0.01784  0.07993
                nmis      fmi   lambda
(Intercept)      NA 0.022932 0.018605
female            0 0.006338 0.002045
i1                0 0.005665 0.001373
sexrisk           1 0.007615 0.003322
indtot           14 0.026436 0.022095
```

The summary includes the number of missing observations as well as the fraction of missing information (fmi). While the results are qualitatively similar, they do differ, which is not surprising given the different imputation models used. Support for other missing data models is available in the `mix` and `mitools` packages.

11.4.5 Finite mixture models with concomitant variables

Finite mixture models (FMMs) can be used in settings where some unmeasured classification separates the observed data into groups with different exposure–outcome relationships. One familiar example of this is a zero-inflated model (7.2.1), where some observations come from a degenerate distribution with all mass at 0. In that case, the exposure–outcome relationship is less interesting in the degenerate distribution group, but there would be considerable interest in the estimated probability of group membership. Another possibly familiar setting is the estimation of a continuous density as a mixture of normal distributions.

More generally, there could be several groups, with "concomitant" covariates predicting group membership. Each group might have different sets of predictors and outcomes from different distribution families. On the other hand, in a "homogenous" mixture setting, all groups have the same distributional form, but with different parameter values. If the covariates in the model are the same, this setting is similar to an ordinary regression model where every observed covariate interacts with the (unobserved) group membership variable.

We'll demonstrate with a simulated dataset. We create a variable x that predicts both group membership and an outcome y with different linear regression parameters depending on group. The mixing probability follows a logistic regression with intercept = −1 and slope (log odds ratio) = 2.

The intercept and slope for the outcome are $(0, 1)$ and $(3, 1.2)$ for the groups, respectively. We leave as an exercise for the reader to explore the consequences of naively fitting the model but ignoring the mixture.

```
> set.seed(1492)
> n = 10000
> x = rnorm(n)
```

```
> probgroup1 = exp(-1 + 2*x)/(1 + exp(-1 + 2*x))
> group = ifelse(probgroup1 > runif(n), 1, 0)
> y = (group * 3) + ((1 + group/5) * x) + rnorm(n)
```

To fit the model, we'll use the `flexmix` package [96], a quite general tool written by Gruen, Leisch, and Sarkar (other options are described in the CRAN finite mixture models task view).

```
> library(flexmix)
> mixout.fm=flexmix(y ~ x, k=2, model=FLXMRglmfix(y ~ x, varFix=TRUE),
    concomitant=FLXPmultinom(~ x))
```

The `flexmix()` function uses a variety of special objects that are created by other functions the package provides. Here, we use the `FLXMRglmfix()` function to force equal variances across the components and the `FLXPmultinom()` function to define the logistic regression on the covariate x for the concomitant model.

The results can be generated from the output object with the `parameters()` function. By default, it prints the parameters for the model in each component.

```
> parameters(mixout.fm)
                  Comp.1     Comp.2
coef.(Intercept)  2.938   -0.009559
coef.x            1.218    0.986896
sigma             1.008    1.008276
```

Using the `which="concomitant"` option generates the parameter estimates for the concomitant model.

```
> parameters(mixout.fm, which="concomitant")
              1        2
(Intercept)   0   0.9767
x             0  -2.0174
```

Impressive accuracy has been achieved.

11.5 Further resources

Comprehensive descriptions of reproducible analysis tools and workflow can be found in [199] and [47], while the `greport` package (due to Frank Harrell) has many powerful features for reporting of clinical trials [83]. Gelman et al. [50] is an accessible introduction to Bayesian inference, while Albert [4] focuses on the use of R for Bayesian computations. Rubin's review [142] and Schafer's book [149] provide overviews of multiple imputation, while [177, 131, 176] describe chained equation models. Review of software implementations of missing data models can be found in [71, 70].

Chapter 12

Case studies

In this chapter, we explore several case studies that demonstrate the statistical computing strengths and potential of R. This includes data management tasks, reading more complex files, creating maps, data scraping, using the `shiny` system within RStudio, manipulating larger datasets, and solving an optimization problem.

12.1 Data management and related tasks

12.1.1 Finding two closest values in a vector

Suppose we need to find the closest pair of observations for some variable. This might arise if we were concerned that some data had been accidentally duplicated. In this case study, we return the IDs of the two closest observations and their distance from each other. We'll first create some sample data and sort it, recognizing that the smallest difference must come between two observations that are adjacent after sorting.

We begin by generating data (3.1.6), along with some subject identifiers (2.3.4).

```
> options(digits=3)
> ds = data.frame(x=rnorm(8), id=1:8)
```

Then, we sort the data. The `order()` function (2.3.10) is used to keep track of the sorted random variables.

```
> options(digits=3)
> ds = ds[order(ds$x),]
> ds
       x id
5 -0.893  5
4 -0.729  4
3 -0.609  3
2 -0.518  2
7 -0.436  7
1  0.180  1
6  1.222  6
8  1.387  8
```

We can use the `diff()` function to get the differences between observations. The `which.min()` function extracts the index (location within the vector) of the smallest value. We apply this function to the `diffx` vector to find the location and extract that location from the `id` vector.

```
> diffx = diff(ds$x)
> min(diffx)
[1] 0.0822
> with(ds, id[which.min(diffx)]) # first val
[1] 2
> with(ds, id[which.min(diffx) + 1]) # second val
[1] 7
```

12.1.2   Tabulate binomial probabilities

Suppose we wanted to assess the probability $P(X = x)$ for a binomial random variate with $n = 10$ and with $p = .81, .84, \ldots, .99$. This could be helpful, for example, in various game settings.

We make a vector of the binomial probabilities, using the : operator (2.3.4) to generate a sequence of integers. After creating an empty matrix (3.3) to hold the table results, we loop (4.1.1) through the binomial probabilities, calling the `dbinom()` function (3.1.1) to find the probability that the random variable takes on that particular value. This calculation is nested within the `round()` function (3.2.4) to reduce the digits displayed. Finally, we include the vector of binomial probabilities with the results using `cbind()`.

```
> p = .78 + (3 * 1:7)/100
> allprobs = matrix(nrow=length(p), ncol=11)
> for (i in 1:length(p)) {
     allprobs[i,] = round(dbinom(0:10, 10, p[i]),2)
  }
> table = cbind(p, allprobs)
```

```
> table
        p
[1,] 0.81 0 0 0 0 0 0.02 0.08 0.19 0.30 0.29 0.12
[2,] 0.84 0 0 0 0 0 0.01 0.05 0.15 0.29 0.33 0.17
[3,] 0.87 0 0 0 0 0 0.00 0.03 0.10 0.25 0.37 0.25
[4,] 0.90 0 0 0 0 0 0.00 0.01 0.06 0.19 0.39 0.35
[5,] 0.93 0 0 0 0 0 0.00 0.00 0.02 0.12 0.36 0.48
[6,] 0.96 0 0 0 0 0 0.00 0.00 0.01 0.05 0.28 0.66
[7,] 0.99 0 0 0 0 0 0.00 0.00 0.00 0.00 0.09 0.90
```

12.1.3   Calculate and plot a running average

The Law of Large Numbers concerns the convergence of the arithmetic average to the expected value, as sample sizes increase. This is an important topic in mathematical statistics. The convergence (or lack thereof, for certain distributions) can easily be visualized [68].

We define a function (4.2) to calculate the running average for a given vector, allowing for variates from many distributions to be generated.

```
> runave = function(n, gendist, ...) {
    x = gendist(n, ...)
    avex = numeric(n)
    for (k in 1:n) {
        avex[k] = mean(x[1:k])
    }
    return(data.frame(x, avex))
}
```

The `runave()` function takes, at a minimum, two arguments: a sample size **n** and function (4.2) denoted by `gendist` that is used to generate samples from a distribution (3.1). In addition, other options for the function can be specified, using the ... syntax (see 4.2). This is used, for example, to specify the degrees of freedom for the samples generated for the t distribution in the next code block. The loop in the `runave()` function could be eliminated through use of the `cumsum()` function applied to the vector given as an argument, and then divided by a vector of observation numbers.

Next, we generate the data, using our new macro and function. To make sure we have a nice example, we first set a fixed seed (3.1.3). Recall that because the expectation of a Cauchy random variable is undefined [137], the sample average does not converge to the center, while a t distribution with more than 1 degree of freedom does.

```
> vals = 1000
> set.seed(1984)
> cauchy = runave(vals, rcauchy)
> t4 = runave(vals, rt, 4)
```

Now we can plot the results. We begin with an empty plot with the correct axis limits, using the `type="n"` specification (8.3.1). We add the running average using the `lines()` function (9.1.1) and varying the line style (9.2.11) and thickness (9.2.12) with the `lty` and `lwd` specifications, respectively. Finally, we specify a title (9.1.9) and a legend (9.1.15). The results are displayed in Figure 12.1.

12.1.4   Create a Fibonacci sequence

The Fibonacci numbers have many mathematical relationships and have been discovered repeatedly in nature. They are constructed as the sum of the previous two values, initialized with the values 1 and 1. It's convenient to use a **for** loop, though other approaches (e.g., recursion) could be used.

```
> len = 10
> fibvals = numeric(len)
> fibvals[1] = 1
> fibvals[2] = 1
> for (i in 3:len) {
    fibvals[i] = fibvals[i-1] + fibvals[i-2]
  }
> fibvals
 [1]  1  1  2  3  5  8 13 21 34 55
```

```
> plot(c(cauchy$avex, t4$avex), xlim=c(1, vals), type="n")
> lines(1:vals, cauchy$avex, lty=1, lwd=2)
> lines(1:vals, t4$avex, lty=2, lwd=2)
> abline(0, 0)
> legend(vals*.6, -1, legend=c("Cauchy", "t with 4 df"),
     lwd=2, lty=c(1, 2))
```

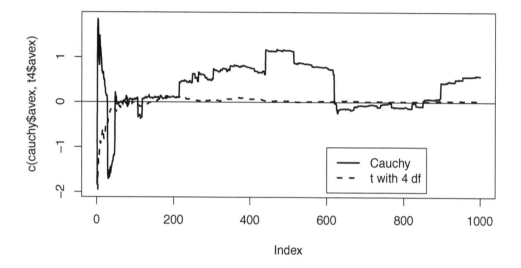

Figure 12.1: Running average for Cauchy and *t* distributions

12.2   Read variable format files

Sometimes datasets are stored in variable format. For example, US Census boundary files (available from http://www.census.gov/geo/www/cob/index.html) are available in both proprietary and ASCII formats. An example ASCII file describing the counties of Massachusetts is available on the book website (http://www.amherst.edu/~nhorton/r2). The first few lines are reproduced here.

```
       1       -0.709816806854972E+02        0.427749187746914E+02
     -0.709148990000000E+02        0.428865890000000E+02
     -0.709148860000000E+02        0.428865640000000E+02
     -0.709148860000000E+02        0.428865640000000E+02
     -0.709027680000000E+02        0.428865300000000E+02
...
     -0.709148990000000E+02        0.428865890000000E+02
END
```

The first line contains an identifier for the county (linked with a county name in an additional file) and a latitude and longitude centroid within the polygon representing the county defined by the remaining points. The remaining points on the boundary do not contain the identifier. After the lines with the points, a line containing the word "END" is included. In addition, the county boundaries contain different numbers of points. The county names, which can be associated by the county identifier, are stored in another dataset. To get the names onto the map, we have to merge the centroid location dataset with the county names dataset. They have to be sorted first. Reading this kind of data requires some care in programming.

We begin by reading in all of the input lines, keeping track of how many counties have been observed (based on how many lines include END). This information is needed for housekeeping purposes when collecting map points for each county.

```
> # read in the data
> input =
  readLines("http://www.amherst.edu/~nhorton/r2/datasets/co25_d00.dat",
    n=-1)
> # figure out how many counties, and how many entries
> num = length(grep("END", input))
> allvals = length(input)
> numentries = allvals-num
> # create vectors to store data
> county = numeric(numentries);
> lat = numeric(numentries)
> long = numeric(numentries)
```

Each line of the input file is processed in turn.

```
> curval = 0    # number of counties seen so far
> # loop through each line
> for (i in 1:allvals) {
    if (input[i]=="END") {
      curval = curval + 1
    } else {
      # remove extraneous spaces
      nospace = gsub("[ ]+", " ", input[i])
      # remove space in first column
      nospace = gsub("^ ", "", nospace)
      splitstring = as.numeric(strsplit(nospace, " ")[[1]])
      len = length(splitstring)
      if (len==3) {  # new county
        curcounty = splitstring[1]; county[i-curval] = curcounty
        lat[i-curval] = splitstring[2]; long[i-curval] = splitstring[3]
      } else if (len==2) { # continue current county
        county[i-curval] = curcounty; lat[i-curval] = splitstring[1]
        long[i-curval] = splitstring[2]
      }
    }
  }
```

The strsplit() function is used to parse the input file. Lines containing END require incrementing the count of counties seen to date. If the line indicates the start of a new county, the new county number is saved. If the line contains two fields (another set of latitudes and longitudes), then this information is stored in the appropriate index (i-curval) of the output vectors.

Next we read in a dataset of county names.

```
> # read county names
> countynames =
  read.table("http://www.amherst.edu/~nhorton/r2/datasets/co25_d00a.dat",
     header=FALSE)
> names(countynames) = c("county", "countyname")
```

12.3   Plotting maps

12.3.1   Massachusetts counties, continued

We're ready to plot the Massachusetts counties and annotate the plot with the names of
the counties.

To make the map, we begin by determining the plotting region, creating the plot of
boundaries, then adding the county names at the internal point that was provided. Since
the first set of points is in the interior of the county, these are not included in the values
given to the polygon function (see indexing, A.4.2).

```
> counties = unique(county)
> xvals = c(min(lat), max(lat)); yvals = c(range(long))
```

The results are displayed in Figure 12.2. Many other maps as well as more sophisticated
projections are supported with the maps package (see also the CRAN spatial statistics task
view).

```
> plot(xvals, yvals, pch=" ", xlab="", ylab="", xaxt="n", yaxt="n")
> for (i in 1:length(counties)) {  # first element is an internal point
    polygon(lat[county==counties[i]][-1], long[county==counties[i]][-1])
    # plot name of county using internal point
    text(lat[county==counties[i]][1], long[county==counties[i]][1],
      countynames$countyname[i], cex=0.8)
  }
```

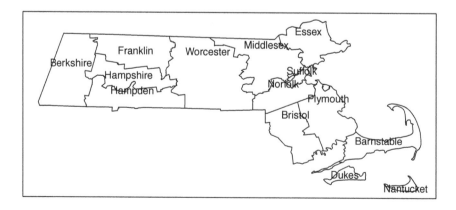

Figure 12.2: Massachusetts counties

12.3.2 Bike ride plot

The Pioneer Valley of Massachusetts, where we both live, is a wonderful place to take a bike ride. In combination with technology to track GPS coordinates, time, and altitude, information regarding outings can be displayed. The data used here can be downloaded, as demonstrated below, from `http://www.amherst.edu/~nhorton/r2/datasets/cycle.csv`.

A map can be downloaded for a particular area from Google Maps, then plotted in conjunction with latitude/longitude coordinates using functions in the `ggmap` package. These routines are built on top of the `ggplot2` "grammar of graphics" package. We found good locations for Amherst using trial and error and plotted the bike ride GPS signals with the map.

```
> library(ggmap)
> options(digits=4)
> amherst = c(lon=-72.52, lat=42.36)
> mymap = get_map(location=amherst, zoom=13, color="bw")
```

```
> myride =
    read.csv("http://www.amherst.edu/~nhorton/r2/datasets/cycle.csv")
> head(myride, 2)
                 Time Ride.Time Ride.Time..secs. Stopped.Time
1 2010-10-02 16:26:54   0:00:01              0.9      0:00:00
2 2010-10-02 16:27:52   0:00:59             58.9      0:00:00
  Stopped.Time..secs. Latitude Longitude Elevation..feet. Distance..miles.
1                   0    42.32    -72.51              201             0.00
2                   0    42.32    -72.51              159             0.04
  Speed..miles.h.    Pace Pace..secs. Average.Speed..miles.h. Average.Pace
1              NA                  NA                     0.00      0:00:00
2            2.73 0:21:56        1316                     2.69      0:22:17
  Average.Pace..secs. Climb..feet. Calories
1                   0            0        0
2                1337            0        1
```

The results are shown in Figure 12.3. Relatively poor cell phone service leads to sparsity in the points in the middle of the figure. For more complex multi-dimensional graphics made with the same data, see `http://tinyurl.com/sasrblog-bikeride` and `http://tinyurl.com/sasrblog-bikeride-redux`.

12.3.3 Choropleth maps

Choropleth maps (see 8.5.1 and 12.6.2) are helpful for visualizing geographic data. In this example, we use data from the built-in R dataset, `USArrests`, which includes United States arrests in 1973 per 100,000 inhabitants in various categories by state.

We'll use the `ggmap` package to generate the plot. It builds on the `ggplot2` package, which implements ideas related to the "grammar of graphics" [188]. The package uses a syntax where specific elements of the plot are added to the final product using special functions connected by the + symbol. Some additional work is needed to merge the dataset with the state information (2.3.11) and to sort the resulting dataframe (2.3.10) so that the shape data for the states is plotted in order.

```
> ggmap(mymap) + geom_point(aes(x=Longitude, y=Latitude), data=myride)
```

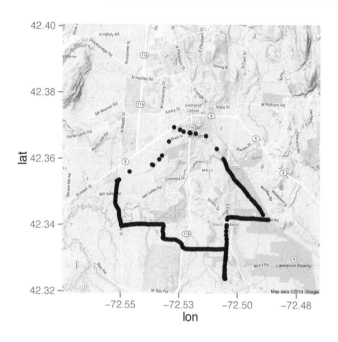

Figure 12.3: Bike ride plot

```
> library(ggmap); library(dplyr)
> USArrests.st = mutate(USArrests,
    region=tolower(rownames(USArrests)),
    murder = cut_number(Murder, 5))
> us_state_map = map_data('state')
> map_data = merge(USArrests.st, us_state_map, by="region")
> map_data = arrange(map_data, order)
> head(select(map_data, region, Murder, murder, long, lat, group, order))
   region Murder       murder  long  lat group order
1 alabama   13.2 (12.1,17.4] -87.5 30.4     1     1
2 alabama   13.2 (12.1,17.4] -87.5 30.4     1     2
3 alabama   13.2 (12.1,17.4] -87.5 30.4     1     3
4 alabama   13.2 (12.1,17.4] -87.5 30.3     1     4
5 alabama   13.2 (12.1,17.4] -87.6 30.3     1     5
6 alabama   13.2 (12.1,17.4] -87.6 30.3     1     6
```

The scale_fill_grey() function changes the colors from the default unordered multiple colors to an ordered and print-friendly black and white (see also scale_file_brewer). The ggmap package uses the Mercator projection (see coord_map() in the ggplot2 package and mapproject in the mapproject package). Another implementation of choropleth maps can be found in the choroplethr package.

The results are displayed in Figure 12.4. As always, the choice of groupings can have an impact on the message conveyed by the graphical display.

```
> p0 = ggplot(map_data, aes(x=long, y=lat, group=group)) +
    geom_polygon(aes(fill = murder)) +
    geom_path(colour='black') +
    theme(legend.position = "bottom",
      panel.background=element_rect(fill="transparent",
        color=NA)) +
    scale_fill_grey(start=1, end =.1) + coord_map();
> plot(p0)
```

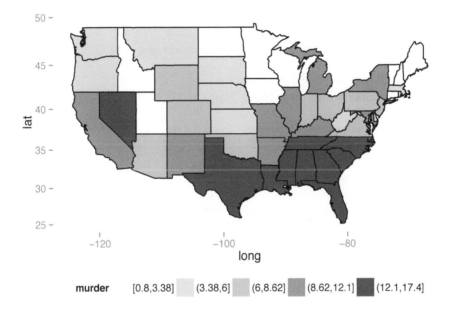

Figure 12.4: Choropleth map

12.4 Data scraping

In this section, we demonstrate various methods for extracting data from web pages, directly from URLs, via APIs, or using table formats.

12.4.1 Scraping data from HTML files

Here, we automate data harvesting from the web, by "scraping" a URL, then reading a datafile with two lines per observation, and plotting the results as time series data. The data being harvested and displayed are the sales ranks from Amazon for the *Cartoon Guide to Statistics* [53].

We can find the Amazon sales rank for a book by downloading the HTML code for a desired web page and searching for the appropriate line. The code to do this relies heavily on 1.1.9 (reading more complex data files) as well as 2.2.14 (replacing strings).

An example can be found at http://www.amherst.edu/~nhorton/r2/datasets/cartoon.html. Many thousands of lines into the file, we find the line we're looking for.

```
#8,048 in Books (<a href="http://www.amazon.com/best-sellers-books-
  Amazon/zgbs/books/ref=pd_dp_ts_b_1">See Top 100 in Books</a>)
```

(We've inserted a line break to allow for printing). If you want to see this, find out how to `View Source` in your browser. In Mozilla Firefox, this is in the `Web Developer` tab.

Unfortunately, the line number where it appears changes periodically. Thus, to find the line, we need to read every line of the file and parse it until we find the correct line. Our approach will be to first look for the line with the expression `<b>See Top 100 in Books</b>`. Once we've found the line, we can look for the `#` symbol, and the numbers between there and the text `in Books`.

We'll use a function to isolate the number, as annotated within the code below. To help in comprehending the code, readers are encouraged to run the commands on a line-by-line basis, then look at the resulting value.

```
> # grab contents of web page
> urlcontents = readLines("http://tinyurl.com/cartoonguide")
>
> # find line with sales rank
> linenum = suppressWarnings(grep("See Top 100 in Books", urlcontents))
>
> # split line into multiple elements
> linevals = strsplit(urlcontents[linenum], ' ')[[1]]
>
> # find element with sales rank number
> entry = grep("#", linevals)
> charrank = linevals[entry] # snag that entry
> charrank = substr(charrank, 2, nchar(charrank)) # kill '#' at start
> charrank = gsub(',' ,'', charrank) # remove commas
> salesrank = as.numeric(charrank) # make it numeric
> cat("salesrank=", salesrank, "\n")
```

In our experience, the format of Amazon's book pages changes often. The code above may not work on current pages, but could be tested on the example page mentioned above, at `http://www.amherst.edu/~nhorton/r2/datasets/cartoon.html`. More sophisticated approaches to web scraping can be found in the `httr` package as well as Nolan and Temple Lang [122].

12.4.2   Reading data with two lines per observation

The code from 12.4.1 was run regularly on a server by calling R in batch mode (see A.2.2), with results stored in a cumulative file. While a date stamp was added, it was included in the file on a different line. The file (accessible at `https://www.amherst.edu/~nhorton/r2/datasets/cartoon.txt`) has the following form.

```
Wed Oct  9 16:00:04 EDT 2013
salesrank= 3269
Wed Oct  9 16:15:02 EDT 2013
salesrank= 4007
```

We begin by reading the file, then we calculate the number of entries by dividing the file's length by two. Next, two empty vectors of the correct length and type are created to store the data. Once this preparatory work is completed, we loop (4.1.1) through the file, reading in the odd-numbered lines as date/time values from the Eastern US time zone, with daylight savings applied. The `gsub()` function (2.2.14) replaces matches determined by regular expression matching. In this situation, it is used to remove the time zone from the line before this processing. These date/time values are read into the `timeval` vector. Even-numbered lines are read into the rank vector, after removing the strings `salesrank=` and `NA` (again using two calls to `gsub()`). Finally, we make a dataframe (A.4.6) from the two vectors and display the first few lines using the `head()` function (1.2.1).

```
> library(RCurl)
> myurl =
    getURL("https://www3.amherst.edu/~nhorton/r2/datasets/cartoon.txt",
            ssl.verifypeer=FALSE)
> file = readLines(textConnection(myurl))
> n = length(file)/2
> rank = numeric(n)
> timeval = as.POSIXlt(rank, origin="1960-01-01")
> for (i in 1:n) {
    timeval[i] = as.POSIXlt(gsub('EST', '',
        gsub('EDT', '', file[(i-1)*2+1])),
        tz="EST5EDT", format="%a %b %d %H:%M:%S %Y")
    rank[i] = as.numeric(gsub('NA', '',
        gsub('salesrank= ','', file[i*2])))
  }
> timerank = data.frame(timeval, rank)
```

Note that the file is being read from an HTTPS (Hypertext Transfer Protocol Secure) connection (1.1.12) and string data is converted to date and time variables (2.4.6). The first four entries of the file are given below.

```
> head(timerank, 4)
            timeval rank
1 2013-09-30 00:00:03 5151
2 2013-09-30 00:15:03 5151
3 2013-09-30 00:30:03 4162
4 2013-09-30 00:45:03 4162
```

12.4.3   Plotting time series data

While it is straightforward to make a simple plot of the data from 12.4.2 using code discussed in 8.3.1, we'll augment the display by indicating whether the rank was recorded in the nighttime (eastern US time) or not. Then we'll color the nighttime ranks differently from the daytime ranks.

We begin by creating a new variable reflecting the date-time at the midnight before we started collecting data. We then coerce the time values to numeric values using the `as.numeric()` function (2.2.7) while subtracting that midnight value. Next, we call the `hour()` function in the `lubridate` package (2.4) to get the hour of measurement.

```
> library(lubridate)
> timeofday = hour(timeval)
> night = rep(0,length(timeofday))  # vector of zeroes
> night[timeofday < 8 | timeofday > 18] = 1
```

The time series plot is requested by the `type="l"` option and symbols for the ranks added with calls to the `points()` function. The `abline()` function adds a reference line at the start of October. The results are displayed in Figure 12.5.

```
> plot(timeval, rank, type="l", xlab="", ylab="Amazon Sales Rank")
> points(timeval[night==1], rank[night==1], cex=0.7, pch=3, col="black")
> points(timeval[night==0], rank[night==0], cex=0.7, pch=4, col="grey")
> legend(as.POSIXlt("2013-10-03 00:00:00 EDT"), 6000,
      legend=c("day","night"), col=c("grey","black"), pch=c(4,3))
> abline(v=as.numeric(as.POSIXlt("2013-10-01 00:00:00 EST")), lty=2)
```

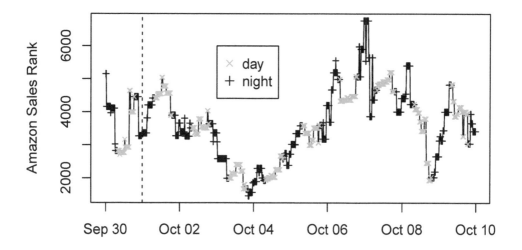

Figure 12.5: Sales plot by time

12.4.4   Reading tables from HTML

In this example, we demonstrate how to read from an HTML table (1.1.14), in this case, the list from Wikipedia of all movies and television shows set in or shot in Liverpool, England. The URL of interest can be found at http://tinyurl.com/liverpoolTV. As of September 2014, the second table consisted of the list of all movies set in or shot in Liverpool. (For future reference, a version of the saved webpage can be found at http://www.amherst.edu/~nhorton/r2/datasets/liverpool.html.)

```
> require(XML)
> require(mosaic)
> wikipedia = "http://en.wikipedia.org/wiki"
> liverpool = "List_of_films_and_television_shows_set_or_shot_in_Liverpool"
> result = readHTMLTable(paste(wikipedia, liverpool, sep="/"),
    stringsAsFactors=FALSE)
> table1 = result[[2]]
> names(table1)
[1] "Title" "Year"  "Notes"
```

We can undertake some data management to reformat and reorganize the data, then display the first records and summary statistics on the year of release.

```
> require(dplyr)
> finaltable = table1 %>%
    mutate(year = as.numeric(Year)) %>%
    select(year, Title)
> head(finaltable, 8)
  year                  Title
1 1901  The Arrest of Goudie
2 1994      Blood on the Dole
3 2009 Charlie Noades R.I.P.
4 2003             Dad's Dead
5 1987      Business as Usual
6 1990 Dancin' Thru the Dark
7 1994            Dark Summer
8 1999      David Copperfield
> favstats(~ year, data=finaltable)
  min   Q1 median   Q3  max mean   sd  n missing
 1901 1986   1994 2001 2010 1988 21.6 40       0
```

The relatively small number of movies facilitates display of the individual values using a stem plot (8.1.2). We note that the number of movies increased dramatically in the 1980s (more than a decade post-Beatles).

```
> with(finaltable, stem(year, scale=2))

  The decimal point is 1 digit(s) to the right of the |

  190 | 1
  191 |
  192 |
  193 | 8
  194 |
  195 | 09
  196 | 59
  197 | 16
  198 | 357888
  199 | 011244445799
  200 | 0011234566689
  201 | 0
```

12.4.5 URL APIs and truly random numbers

Usually, we're content to use a pseudo-random number generator. But sometimes we may want numbers that are actually random. An example might be for randomizing treatment status in a randomized controlled trial. The site `Random.org` provides truly random numbers based on radio static. For long simulations that need a huge number of random numbers, the quota system at `Random.org` may preclude its use. But for small to moderate needs, it can be used to provide truly random numbers. In addition, you can purchase larger quotas if need be.

The site provides application programming interfaces (APIs) for several types of information. We'll demonstrate how to use these to pull vectors of uniform (0,1) random numbers (of 10^{-9} precision) and to check the quota. To generate random variates from other distributions, you can use the inverse probability integral transform (3.1.10).

Two functions are shown below. It is necessary to enclose the character string for the URL in the as.character() function (1.1.9).

```
> truerand = function(numrand) {
      read.table(as.character(paste("http://www.random.org/integers/?num=",
      numrand, "&min=0&max=1000000000&col=1&base=10&format=plain&rnd=new",
      sep="")))/1000000000
  }
>
> quotacheck = function() {
      line = as.numeric(readLines(
          "http://www.random.org/quota/?format=plain"))
      return(line)
  }
```

```
> truerand(7)
     V1
1 0.901
2 0.912
3 0.387
4 0.990
5 0.154
6 0.225
7 0.780
> quotacheck()
[1] 1e+06
```

12.4.6   Reading from a web API

The httr package facilitates access to web application program interfaces (API). It splits the operation into two parts: the request (data sent to the server), and the response (data sent back from the server). As an example, consider a search for items that are tagged as being related to the dplyr package on the stackexchange.com website (see api.stackexchange.com for more information about the interface). Figure 12.6 displays a list of questions that are tagged in this manner.

```
> library(httr)
> # Find the most recent R questions on stackoverflow
> getresult = GET("http://api.stackexchange.com",
>   path="questions",
>   query=list(site="stackoverflow.com", tagged="dplyr"))
> stop_for_status(getresult) # Ensure returned without error
> questions = content(getresult)  # Grab content
```

The GET() function retrieves information from the specified URL: this can be configured to transfer only new information (if data have already been requested). It is always a good practice to check for errors. The content() function is used to extract content from a request, which can then be processed.

Figure 12.6: List of questions tagged with `dplyr` on the Stackexchange website

```
> names(questions$items[[1]])     # What does the returned data look like?
 [1] "tags"                "owner"            "is_answered"
 [4] "view_count"          "answer_count"     "score"
 [7] "last_activity_date"  "creation_date"    "question_id"
[10] "link"                "title"
> substr(questions$items[[1]]$title, 1, 68)
[1] "same function name: possible conflict between dplyr and other packag"
> substr(questions$items[[2]]$title, 1, 68)
[1] "determine observations not included by filtering with dplyr - R"
> substr(questions$items[[3]]$title, 1, 68)
[1] "Using R, how to split a data frame&#39;s column and then break into "
```

Further analysis can be undertaken with the information provided by the API. The package also supports cookies, extraction of status codes, and progress bars (for extended downloads).

12.5   Text mining

12.5.1   Retrieving data from arXiv.org

The `aRxiv` package facilitates access to `arXiv.org`, a repository of electronic preprints for a number of scientific disciplines. It receives many thousands of new submissions each month.

```
> library(aRxiv)
> library(lubridate)
> library(stringr)
> library(dplyr)
> efron = arxiv_search(query='au:"Efron" AND cat:stat*', limit=50)
> names(efron)
 [1] "id"             "submitted"        "updated"
 [4] "title"          "abstract"         "authors"
 [7] "affiliations"   "link_abstract"    "link_pdf"
[10] "link_doi"       "comment"          "journal_ref"
[13] "doi"            "primary_category" "categories"
> dim(efron)
[1] 14 15
> efron = mutate(efron, submityear =
    year(sapply(str_split(submitted, " "), "[[", 1)))
> with(efron, table(submityear))
submityear
2004 2006 2007 2008 2009 2010 2013 2014
   2    1    1    4    1    2    2    1
```

In this example, submissions from eminent statistician Bradley Efron are downloaded, including submission dates, lists of authors, titles, keywords, and abstracts. Note that the double quotes are nested within the single quotes. Functions from the `lubridate` and `stringr` packages facilitate processing the date and time string (e.g., `"2004-06-23 12:59:32"` for "Rejoinder to 'Least angle regression'" from *Annals of Statistics*). The string is split into a date and time (using `str_split()`), turned into a vector (using `sapply()`), then the year value is extracted (using `year()`). The distribution of the 14 papers can be displayed as a table.

Care should be taken not to overload the `arXiv` server. The `arxiv_count()` function should be run to determine the number of matches for a given search, to allow larger requests to be downloaded in chunks.

12.5.2   Exploratory text mining

Text mining (a form of text analytics) is a fast-growing application of statistical and machine learning techniques. There are a number of R packages that facilitate analysis of text documents. In this example, we utilize functions from the `tm` package to generate a corpus of documents (a structured set of texts) consisting of the abstracts from the previous search of papers on `arxiv.org` by Brad Efron (see 12.5.1).

Here, we use the `DataframeSource()` function to create the corpus. Other possible sources of text include directories, XML, or URIs (universal resource identifiers). We then display the first abstract.

```
> library(tm)
> mycorpus = VCorpus(DataframeSource(data.frame(efron$abstract)))
> head(strwrap(mycorpus[[1]]))
[1] "The purpose of model selection algorithms such as All Subsets,"
[2] "Forward Selection and Backward Elimination is to choose a linear"
[3] "model on the basis of the same set of data to which the model will"
[4] "be applied. Typically we have available a large collection of"
[5] "possible covariates from which we hope to select a parsimonious"
[6] "set for the efficient prediction of a response variable. Least"
```

Next we want to clean up the corpus. The `tm_map()` function is called repeatedly to strip whitespace, remove numbers and punctuation, map all of the text to lowercase, and elide common English words.

```
> mycorpus = tm_map(mycorpus, stripWhitespace)
> mycorpus = tm_map(mycorpus, removeNumbers)
> mycorpus = tm_map(mycorpus, removePunctuation)
> mycorpus = tm_map(mycorpus, content_transformer(tolower))
> mycorpus = tm_map(mycorpus, removeWords, stopwords("english"))
> head(strwrap(mycorpus[[1]]))
[1] "purpose model selection algorithms subsets forward selection"
[2] "backward elimination choose linear model basis set data model will"
[3] "applied typically available large collection possible covariates"
[4] "hope select parsimonious set efficient prediction response"
[5] "variable least angle regression lars new model selection algorithm"
[6] "useful less greedy version traditional forward selection methods"
```

Finally, the `DocumentTermMatrix()` can be used to generate a document term matrix. This is a sparse matrix that describes the frequency of terms in a corpus. We can display the terms that arise in 7 or more of the abstracts.

```
> dtm = DocumentTermMatrix(mycorpus)
> findFreqTerms(dtm, 7)
[1] "bayes"        "bayesian"    "empirical"   "evidence"    "frequentist"
[6] "hypothesis"   "methods"     "model"       "new"
```

Many more options for analysis are available (see the CRAN natural language processing task view).

12.6 Interactive visualization

Graphical displays are increasingly interactive, with real-time response to input. A number of systems are available to create such displays within R, including the ggvis package and Shiny. In this section, we will describe these systems and provide examples of their use.

12.6.1 Visualization using the grammar of graphics (ggvis)

The ggvis package provides an interactive "grammar of graphics" [197] to allow web graphics to be displayed and manipulated. It utilizes a syntax similar to the ggplot2 package to create displays, which can be viewed in a browser internally and externally.

The goal is to combine the best of R (e.g., every modeling function you can imagine) and the best of the web (everyone has a web browser). Data manipulation and transformation are done in R and the graphics are then rendered in a web browser. For RStudio users, ggvis graphics display in a viewer panel.

In this example, we create an interactive graphical display using the HELP dataset, where the user can select the size of the points, the opacity (to address overplotting, see 8.3.4), and the fill color.

```
library(ggvis)
> ds = read.csv("http://www.amherst.edu/~nhorton/r2/datasets/help.csv")
> ds %>%
  ggvis(x = ~ mcs, y = ~ cesd,
    size := input_slider(min=10, max=100, label="size"),
    opacity := input_slider(min=0, max=1, label="opacity"),
    fill := input_select(choices=c("red", "green", "blue", "grey"),
      selected="red", label="fill color"),
    stroke := "black") %>%
  layer_points()
```

When the ggvis() and layer_points() functions are run, a viewer window is created (see Figure 12.7). The user is able to adjust each of the controls. More information about ggvis can be found at http://ggvis.rstudio.com.

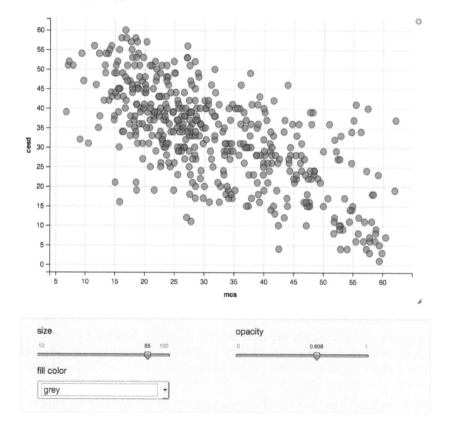

Figure 12.7: Interactive graphical display

12.6.2 Shiny in Markdown

RStudio supports the Shiny system, which is designed to simplify the creation of interactive web applications. It provides automatic "reactive" linkage between inputs and outputs: when the user clicks on one of the radio buttons, sliders, or selections, the output is re-rendered.

Available control widgets include the functions `actionButton()`, `checkboxGroupInput()`, `checkboxInput()`, `dateInput()`, `dateRangeInput()`, `fileInput()`, `helpText()`, `numericInput()`, `radioButtons()`, `selectInput()`, `sliderInput()`, `submitButton()`, and `textInput()`. We demonstrate use of this system by creating an interactive choropleth map of the murder rate in US states (as previously described in 12.3.3). A template can be created by selecting a new Markdown file with the **Shiny** option picked. Figure 12.8 displays a Markdown file that creates a choropleth map that allows control over the number of bins as well as whether to include names of the states.

```
---
title: "Sample Shiny in Markdown"
output: html_document
runtime: shiny
---

Shiny inputs and outputs can be embedded in a Markdown document.  Outputs
are automatically updated whenever inputs change.  This demonstrates
how a standard R plot can be made interactive by wrapping it in the
Shiny 'renderPlot' function. The 'selectInput' function creates the
input widgets used to control the plot display.

'''{r, echo=FALSE}
inputPanel(
  selectInput("n_breaks", label = "Number of breaks:",
    choices = c(2, 3, 4, 5, 9), selected = 5),

  selectInput("labels", label = "Display labels?:",
    choices = c("TRUE", "FALSE"), selected = "TRUE")
)

renderPlot({
  library(choroplethr); library(dplyr)
  USArrests.st = mutate(USArrests,
    region=tolower(rownames(USArrests)),
    value = Murder)
  choroplethr(USArrests.st, "state", title="Murder Rates by State",
    showLabels=input$labels,
    num_buckets=as.numeric(input$n_breaks))
})
```

Figure 12.8: Shiny within R Markdown

The `inputPanel()` function is used in conjunction with the `selectInput()` function to create two widgets: one to control the number of groups and the other to control whether to display the labels for the states.

The `renderPlot()` function can then access these values through the `input` object. The `choroplethr()` function in the `choropleth` package is used to generate the desired figure.

When the document is run (by clicking `Run Document` within RStudio), the results are displayed in a viewer window (see Figure 12.9).

Figure 12.9: Display of Shiny document within Markdown

More information about Shiny can be found at `shiny.rstudio.com`.

12.6.3   Creating a standalone Shiny app

It is possible to create standalone Shiny applications that can be made accessible from the Internet. This has a major advantage over other web application frameworks that require knowledge of HTML, CSS, or JavaScript.

A Shiny application consists of a directory with a file called `app.R` which contains the user-interface definition, server script, and any additional required data, scripts, or other resources. In this example, we will re-create our choropleth plot in a directory in **ShinyApps** called `choropleth`.

```
> library(shiny)
> ui = shinyUI(bootstrapPage(
    selectInput("n_breaks", label="Number of breaks:",
      choices=c(2, 3, 4, 5, 9), selected=5),
    selectInput("labels", label="Display labels?:",
      choices = c("TRUE", "FALSE"), selected="TRUE"),
    plotOutput(outputId="main_plot", height="300px", width="500px")
))
```

```
> server = function(input, output) {
    output$main_plot = renderPlot({
      library(choroplethr); library(dplyr)
      USArrests.st = mutate(USArrests,
        region=tolower(rownames(USArrests)), value = Murder)
      choroplethr(USArrests.st, "state", title="Murder Rates by State",
        showLabels=input$labels, num_buckets=as.numeric(input$n_breaks))
    })
  }
> shinyApp(ui=ui, server=server)
```

The user interface is defined and saved in an object called `ui` by calling the function `bootstrapPage()` and passing the result to the `shinyUI()` function. This defines two selector widgets (through calls to `selectInput()`) and a call to `plotOutput()` (to display the results).

Next we define the server (which is saved in an object called `server`). This utilizes similar code to that introduced in 12.6.2. This process involves creating a function that calls `renderPlot()` after creating the choropleth map.

Finally, the `shinyApp()` function is called to run the app. These commands are all saved in the `app.R` file. The app can be run from within RStudio using the `runApp()` command.

```
> library(shiny)
> runApp("~/ShinyApps/choropleth")
```

The application will then appear in a browser. For those with a Shiny server, the app can be viewed externally (in this case as `https://r.amherst.edu/apps/nhorton/choropleth`). More information about Shiny and Shiny servers can be found at `shiny.rstudio.com`.

12.7 Manipulating bigger datasets

In this example, we consider analysis of the Data Expo 2009 commercial airline flight dataset [189], which includes details of $n = 123,534,969$ flights from 1987 to 2008. We consider the number of flights originating from Bradley International Airport (code BDL, serving Hartford, CT and Springfield, MA). Because of the size of the data, we will demonstrate use of a database system accessed using a structured query language (SQL) [165].

Full details are available on the Data Expo website (`http://stat-computing.org/dataexpo/2009/sqlite.html`) regarding how to download the Expo data as comma-separated files (1.6 gigabytes of compressed, 12 gigabytes uncompressed through 2008), set up and index a database (19 gigabytes), then access it from within R.

A simple way to access databases from R is through SQLite, a self-contained, serverless, transactional SQL database engine. To use this, the analyst installs the `sqlite` software library (`http://sqlite.org`). Next the input files must be downloaded to the local machine, a database set up (by running `sqlite3 ontime.sqlite3`) at the shell command line), table created with the appropriate fields, the files loaded using a series of `.import` statements, and access speeded up by adding indexing. Then the `RSQLite` package can be used to create a connection to the database.

```
> library(RSQLite)
> con = dbConnect("SQLite", dbname = "/Home/Airlines/ontime.sqlite3")
> ds = dbGetQuery(con, "SELECT DayofMonth, Month, Year, Origin,
>    sum(1) as numFlights FROM ontime WHERE Origin='BDL'
>    GROUP BY DayofMonth,Month,Year")
> # returns a dataframe with 7,763 rows and 5 columns
```

The `dbGetQuery()` function in the `RSQLite` package allows an SQL query to be sent to the connection. Here, the SQL statement specifies the five variables to be included (one of which is the count of flights), the name of the table `ontime`, what flights to include (only those originating at BDL), and what level to aggregate (unique day). This dataset can then be post-processed using functions from the `dplyr` package.

```
> library(dplyr)
> ds = mutate(ds, date =
>    as.Date(paste(Year, "-", Month, "-", DayofMonth, sep="")))
> ds = mutate(ds, weekday = weekdays(date))
> ds = arrange(ds, date)
> mondays = filter(ds, weekday=="Monday")
```

The results are plotted in Figure 12.10. Similar functionality is provided for MySQL databases using the `RMySQL` package.

One disadvantage of using SQL is that the syntax is similar to but not equivalent to that of R (for example, a single equal sign is used for comparisons in SQL, but two equal signs in R). The `dplyr` package provides an efficient interface to SQL databases using R syntax. The following code yields the same results as above.

```
> library(dplyr)
> my_db = src_sqlite("/Home/Airlines/ontime.sqlite3")
> my_tbl = group_by(tbl(my_db, "ontime"), DayofMonth, Month, Year, Origin)
> ds = my_tbl %>%
  filter(Origin=="BDL") %>%
  select(DayofMonth, Month, Year, Origin) %>%
  summarise(numFlights=n())
```

12.8   Constrained optimization: the knapsack problem

The website `http://rosettacode.org/wiki/Knapsack_problem/Unbounded` describes a fanciful trip by a traveler to Shangri La. Upon leaving, the traveler is allowed to take as much of three valuable items as they like, as long as they fit in a knapsack. A maximum of 25 weights can be taken, with a total volume of 25 cubic units. The weights, volumes, and values of the three items are given in Table 12.1.

How can the traveler maximize the value of the items? It is straightforward to calculate the solutions using brute force, by iterating over all possible combinations and eliminating those that are overweight or too large to fit.

We define a number of support functions, then run over all possible values of the knapsack contents (after `expand.grid()` generates the list). The `findvalue()` function checks the constraints and sets the value to 0 if they are not satisfied, and otherwise calculates them for the set. The `apply()` function (see 2.6.4) is used to run a function for each item of a vector.

```
> library(lattice)
> xyplot(numFlights ~ date, xlab="", ylab="number of flights on Monday",
>   type="l", col="black", lwd=2, data=mondays)
```

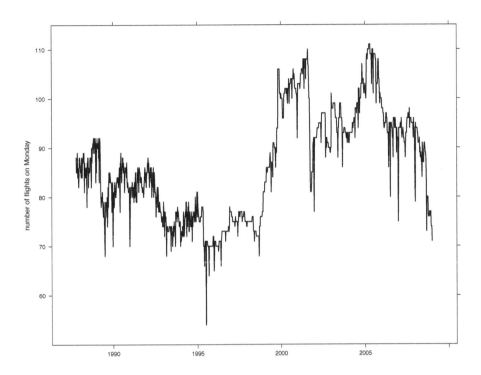

Figure 12.10: Number of flights departing Bradley airport on Mondays over time

```
> # Define constants and useful functions
> weight = c(0.3, 0.2, 2.0)
> volume = c(2.5, 1.5, 0.2)
> value = c(3000, 1800, 2500)
> maxwt = 25
> maxvol = 25
```

```
> # minimize the grid points we need to calculate
> max.items = floor(pmin(maxwt/weight, maxvol/volume))
>
```

Table 12.1: Weights, volume, and values for the knapsack problem

Item	Weight	Volume	Value
Panacea	0.3	2.5	3000
Ichor	0.2	1.5	1800
Gold	2.0	0.2	2500

```
> # useful functions
> getvalue = function(n) sum(n*value)
> getweight = function(n) sum(n*weight)
> getvolume = function(n) sum(n*volume)
>
> # main function: return 0 if constraints not met,
> # otherwise return the value of the contents, and their weight
> findvalue = function(x) {
     thisweight = apply(x, 1, getweight)
     thisvolume = apply(x, 1, getvolume)
     fits = (thisweight <= maxwt) &
           (thisvolume <= maxvol)
     vals = apply(x, 1, getvalue)
     return(data.frame(panacea=x[,1], ichor=x[,2], gold=x[,3],
        value=fits*vals, weight=thisweight,
        volume=thisvolume))
  }
>
> # Find and evaluate all possible combinations
> combs = expand.grid(lapply(max.items, function(n) seq.int(0, n)))
> values = findvalue(combs)
```

Now we can display the solutions.

```
> max(values$value)
[1] 54500
> values[values$value==max(values$value),]
     panacea ichor gold value weight volume
2067       9     0   11 54500   24.7   24.7
2119       6     5   11 54500   24.8   24.7
2171       3    10   11 54500   24.9   24.7
2223       0    15   11 54500   25.0   24.7
```

The first solution (with 9 panacea, no ichor, and 11 gold) satisfies the volume constraint, maximizes the value, and also minimizes the weight. More sophisticated approaches are available using the lpSolve package for linear/integer problems.

Appendix A

Introduction to R and RStudio

This chapter provides a (brief) introduction to R and RStudio. R is a free, open-source software environment for statistical computing and graphics [77, 130]. RStudio is an open-source integrated developement environment for R that adds many features and productivity tools for R. The chapter includes a short history, installation information, a sample session, background on fundamental structures and actions, information about help and documentation, and other important topics.

R is a general-purpose package that includes support for a wide variety of modern statistical and graphical methods (many of which have been contributed by users). It is available for most UNIX platforms, Windows, and MacOS. The R Foundation for Statistical Computing holds and administers the copyright of R software and documentation. R is available under the terms of the Free Software Foundation's GNU General Public License in source code form.

RStudio facilitates use of R by integrating R help and documentation, providing a workspace browser and data viewer, and supporting syntax highlighting, code completion, and smart indentation. It integrates reproducible analysis with Sweave, knitr, and R Markdown (see 11.3), supports the creation of slide presentations, and includes a debugging environment (see 4.1.6). It facilitates the creation of dynamic web applications using Shiny (see 12.6.2). It also provides support for multiple projects as well as an interface to source code control systems such as GitHub. It has become the default interface for many R users, including the authors.

RStudio is available as a client (standalone) for Windows, Mac OS X, and Linux. There is also a server version. Commercial products and support are available in addition to the open-source offerings (see http://www.rstudio.com/ide for details).

The first versions of R were written by Ross Ihaka and Robert Gentleman at the University of Auckland, New Zealand, while current development is coordinated by the R Development Core Team, a group of international volunteers. As of October 2014, this group consisted of Douglas Bates, John Chambers, Peter Dalgaard, Seth Falcon, Robert Gentleman, Kurt Hornik, Ross Ihaka, Michael Lawrence, Friedrich Leisch, Uwe Ligges, Thomas Lumley, Martin Maechler, Martin Morgan, Duncan Murdoch, Paul Murrell, Martyn Plummer, Brian Ripley, Deepayan Sarkar, Duncan Temple Lang, Luke Tierney, and Simon Urbanek. Former members of the R Core include Heiner Schwarte (through 1999), Guido Masarotto (through 2003), and Stefano Iacus (through 2014). Many hundreds of other people have contributed to the development of R or developed add-on libraries and packages.

R is similar to the S language, a flexible and extensible statistical environment originally developed in the 1980s at AT&T Bell Labs (now Alcatel–Lucent). Insightful Corporation has continued the development of S in their commercial software package S-PLUS^{TM}.

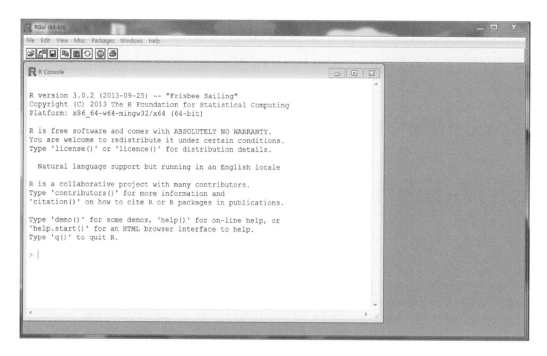

Figure A.1: R Windows graphical user interface

New users are encouraged to download and install R from the Comprehensive R archive network (CRAN, http://www.r-project.org, see A.1) and install RStudio from http://www.rstudio.com/ide. The sample session in the appendix of the *Introduction to R* document, also available from CRAN (see A.2), is highly recommended.

A.1   Installation

The home page for the R project, located at http://r-project.org, is the best starting place for information about the software. It includes links to CRAN, which features pre-compiled binaries as well as source code for R, add-on packages, documentation (including manuals, frequently asked questions, and the R newsletter) as well as general background information. Mirrored CRAN sites with identical copies of these files exist all around the world. Updates to R and packages are regularly posted on CRAN. In addition to the instructions for installation under Windows and Mac OS X, R and RStudio are also available for multiple Linux implementations.

A.1.1   Installation under Windows

Versions of R for Windows XP and later, including 64-bit versions, are available at CRAN. The distribution includes Rgui.exe, which launches a self-contained windowing system that includes a command-line interface, Rterm.exe for a command-line interface only, Rscript.exe for batch processing only, and R.exe, which is suitable for batch or command-line use. A screenshot of the R graphical user interface (GUI) can be found in Figure A.1. More information on Windows-specific issues can be found in the CRAN *R for Windows FAQ* (http://cran.r-project.org/bin/windows/base/rw-FAQ.html).

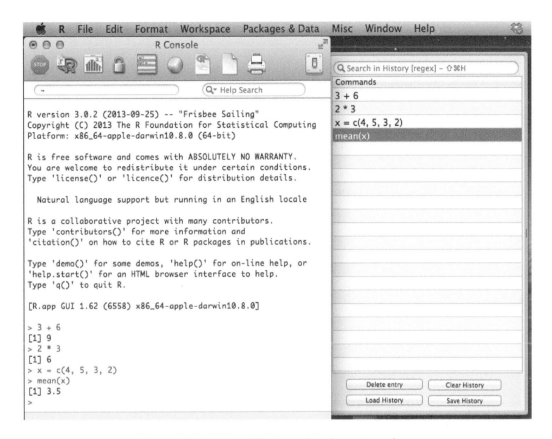

Figure A.2: R Mac OS X graphical user interface

A.1.2 Installation under Mac OS X

A version of R for Mac OS X 10.6 and higher is available at CRAN. This is distributed as a disk image containing the installer. In addition to the graphical interface version, a command-line version (particularly useful for batch operations) can be run as the command R. A screenshot of the graphical interface can be found in Figure A.2.

More information on Macintosh-specific issues can be found in the CRAN *R for Mac OS X FAQ* (http://cran.r-project.org/bin/macosx/RMacOSX-FAQ.html).

A.1.3 RStudio

RStudio for MacOS, Windows, or Linux can be downloaded from http://www.rstudio.com/ide. RStudio requires R to be installed on the local machine. A server version (accessible from web browsers) is also available for download. Documentation of the advanced features in the system is available on the RStudio website. A screenshot of the RStudio interface can be found in Figure A.3.

A.1.4 Other graphical interfaces

Other graphical user interfaces for R include the R Commander project [43], Deducer (http://www.deducer.org), and the SOCR (Statistics Online Computational Resource) project (http://www.socr.ucla.edu).

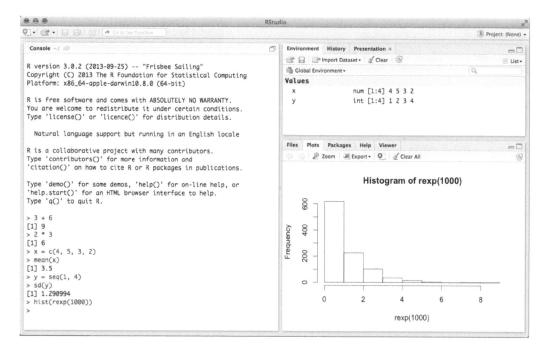

Figure A.3: RStudio graphical user interface

A.2   Running R and sample session

Once installation is complete, the recommended next step for a new user would be to start
R and run a sample session. An example from the command-line interface within Mac OS
X is given in Figure A.4.

The ">" character is the command prompt, and commands are executed once the user
presses the RETURN or ENTER key. R can be used as a calculator (as seen from the
first two commands on lines 1 and 3). New variables can be created (as on lines 5 and 8)
using the assignment operator =. If a command generates output (as on lines 6 and 11),
then it is printed on the screen, preceded by a number indicating place in the vector (this
is particularly useful if output is longer than one line, e.g., lines 23–24). Saved data (here
assigned the name ds) is read into R on line 15, then summary statistics are calculated
(lines 16–17) and individual observations are displayed (lines 23–24). The $ operator allows
access to objects within a dataframe. Alternatively, the with() function can be used to
access objects within a dataset.

It is important to remember that R is case-sensitive.

```
> x = 1:3
> X = seq(2, 4)
> x
[1] 1 2 3
> X
[1] 2 3 4
```

A very comprehensive sample session in R can be found in Appendix A of "An Introduction
to R" [180] (http://cran.r-project.org/doc/manuals/R-intro.pdf).

```
% R
R version 3.1.1 (2014-07-10) -- "Sock it to Me"
Copyright (C) 2014 The R Foundation for Statistical Computing
Platform: x86_64-apple-darwin13.1.0 (64-bit)

R is free software and comes with ABSOLUTELY NO WARRANTY.
You are welcome to redistribute it under certain conditions.
Type 'license()' or 'licence()' for distribution details.

 Natural language support but running in an English locale

R is a collaborative project with many contributors.
Type 'contributors()' for more information and
'citation()' on how to cite R or R packages in publications.

Type 'demo()' for some demos, 'help()' for on-line help, or
'help.start()' for an HTML browser interface to help.
Type 'q()' to quit R.

 1   > 3 + 6
 2   [1] 9
 3   > 2 * 3
 4   [1] 6
 5   > x = c(4, 5, 3, 2)
 6   > x
 7   [1] 4 5 3 2
 8   > y = seq(1, 4)
 9   > y
10   [1] 1 2 3 4
11   > mean(x)
12   [1] 3.5
13   > sd(y)
14   [1] 1.290994
15   > ds = read.csv("http://www.amherst.edu/~nhorton/r2/datasets/help.csv")
16   > mean(ds$age)
17   [1] 35.65342
18   > mean(age)
19   Error in mean(age) : object "age" not found
20   > with(ds, mean(age))
21   [1] 35.65342
22   > ds$age[1:30]
23   [1] 37 37 26 39 32 47 49 28 50 39 34 58 53 58 60 36 28 35 29 27 27
24   [22] 41 33 34 31 39 48 34 32 35
25   > q()
26   Save workspace image? [y/n/c]: n
```

Figure A.4: Sample session in R

A.2.1 Replicating examples from the book and sourcing commands

To help facilitate reproducibility, R commands can be bundled into a plain text file, called a "script" file, which can be executed using the source() command. The optional argument echo=TRUE for the source() command can be set to display each command and its output. The book website cited above includes the R source code for the examples. The sample session in Figure A.4 can be executed by running the following command.

```
> source("http://www.amherst.edu/~nhorton/r2/examples/sampsess.R",
    echo=TRUE)
```

Most of the examples at the end of each chapter can be executed by running the command:

```
> source("http://www.amherst.edu/~nhorton/r2/examples/chapterXX.R",
    echo=TRUE)
```

where XX is replaced by the desired chapter number. In many cases, add-on packages (see A.6.1) need to be installed prior to running the examples. To facilitate this process, we have created a script file to load them in one step.

```
> source("http://www.amherst.edu/~nhorton/r2/examples/install.R",
    echo=TRUE)
```

If needed libraries are not installed (A.6.1), the example code will generate error messages.

A.2.2 Batch mode

In addition, R can be run in batch (noninteractive) mode from a command-line interface:

```
% R CMD BATCH file.R
```

This will run the commands contained within `file.R` and put all output into `file.Rout`. To use R in batch mode under Windows, users need to include R.exe in their path (see the Windows R FAQ and A.1.1).

A.3 Learning R

An excellent starting point for new R users can be found in the *Introduction to R*, available from CRAN (`r-project.org`).

A.3.1 Getting help

The system features extensive online documentation, though it can sometimes be challenging to comprehend. Each command has an associated help file that describes usage, lists arguments, provides details of actions, references, lists other related functions, and includes examples of its use. The help system is invoked using the command:

```
> ?function
```

or

```
> help(function)
```

where `function` is the name of the function of interest. As an example, the help file for the `mean()` function is accessed by the command `help(mean)`. The output from this command is provided in Figure A.5.

It describes the `mean()` function as a generic function for the (trimmed) arithmetic mean, with arguments `x` (an R object), `trim` (the fraction of observations to trim, with default=0; setting `trim=0.5` is equivalent to calculating the median), and `na.rm` (should missing values be deleted; default is `na.rm=F`).

Some commands (e.g., `if`) are reserved, so `?if` will not generate the desired documentation. Running `?"if"` will work (see also `?Reserved` and `?Control`). Other reserved words include `else`, `repeat`, `while`, `function`, `for`, `in`, `next`, `break`, `TRUE`, `FALSE`, `NULL`, `Inf`, `NaN`, and `NA`.

The `RSiteSearch()` function will search for key words or phrases in many places (including the search engine at `http://search.r-project.org`). A screenshot of the results of the command `RSiteSearch("eta squared anova")` can be found in Figure A.6. The `RSeek.org` site can also be helpful in finding more information and examples.

Examples of many functions are available using the `example()` function.

```
> example(mean)

mean> x <- c(0:10, 50)

mean> xm <- mean(x)

mean> c(xm, mean(x, trim = 0.10))
[1] 8.75 5.50
```

Other useful resources are `help.start()`, which provides a set of online manuals, and `help.search()`, which can be used to look up entries by description. The `apropos()` command returns any functions in the current search list that match a given pattern (which facilitates searching for a function based on what it does, as opposed to its name).

Other resources for help available from CRAN include the R-help mailing list (see also A.7, support). The StackOverflow site for R (`http://stackoverflow.com/questions/tagged/r`) provides a series of questions and answers for common questions that are tagged as being related to R. New users are also encouraged to read the R FAQ (frequently asked questions) list. RStudio provides a curated guide to resources for learning R and its extensions (see `http://www.rstudio.com/resources/training/online-learning`).

A.3.2 swirl

The `swirl` system is a collection of interactive courses to teach R programming and data science within the R console (`swirlstats.com`). It requires the installation of the `swirl` package, then use of the `install_from_swirl()` function to download courses. Table A.1 displays the courses that were available as of October 2014. A sample session is displayed below.

```
> library(swirl)

| Type swirl() when you are ready to begin.
```

```
mean                     package:base                    R Documentation

Arithmetic Mean

Description:
     Generic function for the (trimmed) arithmetic mean.

Usage:
     mean(x, ...)

     ## Default S3 method:
     mean(x, trim = 0, na.rm = FALSE, ...)

Arguments:
        x: An R object.  Currently there are methods for numeric/logical
           vectors and date, date-time and time interval objects.
           Complex vectors are allowed for 'trim = 0', only.

     trim: the fraction (0 to 0.5) of observations to be trimmed from
           each end of 'x' before the mean is computed. Values of trim
           outside that range are taken as the nearest endpoint.

    na.rm: a logical value indicating whether 'NA' values should be
           stripped before the computation proceeds.

      ...: further arguments passed to or from other methods.

Value:
     If 'trim' is zero (the default), the arithmetic mean of the values
     in 'x' is computed, as a numeric or complex vector of length one.
     If 'x' is not logical (coerced to numeric), numeric (including
     integer) or complex, 'NA_real_' is returned, with a warning.

     If 'trim' is non-zero, a symmetrically trimmed mean is computed
     with a fraction of 'trim' observations deleted from each end
     before the mean is computed.

References:
     Becker, R. A., Chambers, J. M. and Wilks, A. R. (1988) _The New S
     Language_. Wadsworth & Brooks/Cole.

See Also:
     'weighted.mean', 'mean.POSIXct', 'colMeans' for row and column
     means.

Examples:
     x <- c(0:10, 50)
     xm <- mean(x)
     c(xm, mean(x, trim = 0.10))
```

Figure A.5: Documentation on the mean() function

R Site Search

Query: [eta squared anova] [Search] [How to search]

Display: [20 ÷] **Description:** [normal ÷] **Sort:** [by score   ÷]

Target:
- ☑ Functions
- ☑ Vignettes
- ☑ Task views

For problems WITH THIS PAGE (not with R) contact baron@psych.upenn.edu.

Results:

References:

- **views**: [eta: 1] [squared: 2] [anova: 7] [TOTAL: 0]
- **vignettes**: [eta: 49] [squared: 406] [anova: 193] [TOTAL: 5]
- **functions**: [eta: 962] [squared: 3358] [anova: 1992] [TOTAL: 14]

Total 19 documents matching your query.

1. **R: Measures of Partial Association (Eta-squared) for Linear...** (score: 28)
   Author: *unknown*
   Date: *Thu, 05 Sep 2013 10:24:33 -0500*
   Measures of Partial Association (Eta-squared) for Linear Models Description Usage Arguments Details Value Author(s) References See Also Examples page for etasq {heplots} etasq {heplots} R Documentat
   http://finzi.psych.upenn.edu/R/library/heplots/html/etasq.html (4,148 bytes)

2. **R: Make nice ANOVA table for printing.** (score: 9)
   Author: *unknown*
   Date: *Sat, 07 Dec 2013 07:14:08 -0500*
   Make nice ANOVA table for printing. Description Usage Arguments Details Value Author(s) References See Also Examples page for nice.anova {afex} nice.anova {afex} R Documentation These functions prod
   http://finzi.psych.upenn.edu/R/library/afex/html/nice.anova.html (6,554 bytes)

3. **R: Effect size calculations for ANOVAs** (score: 6)
   Author: *unknown*
   Date: *Wed, 20 Nov 2013 08:27:15 -0500*
   Effect size calculations for ANOVAs Description Usage Arguments Details Value Warning Author(s) See Also Examples page for etaSquared {lsr} etaSquared {lsr} R Documentation Calculates eta-squared an
   http://finzi.psych.upenn.edu/R/library/lsr/html/etaSquared.html (3,770 bytes)

Figure A.6: Display after running `RSiteSearch("eta squared anova")`

Table A.1: Interactive courses available within swirl

COURSE	DESCRIPTION
R Programming (beginner)	The basics of programming in R
R Programming Alt (beginner)	Same as the original, but modified for in-class use
Data Analysis (beginner)	Basic ideas in statistics and data visualization
Mathematical Biostatistics Boot Camp (beginner)	One- and two-sample *t*-tests, power, and sample size
Open Intro (beginner)	A very basic introduction to statistics, data analysis, and data visualization
Regression Models (intermediate)	The basics of regression modeling in R
Getting and Cleaning Data (advanced)	dplyr, tidyr, lubridate, oh my!

```
> install_from_swirl("Getting and Cleaning Data")

| Course installed successfully!

> swirl()

| Welcome to swirl!

| Please sign in. If you've been here before, use the same name as you did
| then. If you are new, call yourself something unique.

What shall I call you? Nick

| Please choose a course, or type 0 to exit swirl.

1: Getting and Cleaning Data
2: R Programming
3: Regression Models
4: Take me to the swirl course repository!

Selection: 1

| Please choose a lesson, or type 0 to return to course menu.

1: Manipulating Data with dplyr
2: Grouping and Chaining with dplyr
3: Tidying Data with tidyr
4: Dates and Times with lubridate

Selection: 1

| Attempting to load lesson dependencies...

| Package dplyr loaded correctly!

| In this lesson, you'll learn how to manipulate data using dplyr. dplyr is
| a fast and powerful R package written by Hadley Wickham and Romain
| Francois that provides a consistent and concise grammar for manipulating
| tabular data.

...
```

After some preliminary introductions, the user is instructed to enter a series of commands
and explore in the console. The `swirl` system detects whether the correct commands have
been input.

A.4 Fundamental structures and objects

Here we provide a brief introduction to R data structures.

A.4.1 Objects and vectors

Almost everything in R is an object, which may be initially disconcerting to a new user. An object is simply something on which R can operate. Common objects include vectors, matrices, arrays, factors (see 2.2.19), dataframes (akin to datasets in other systems), lists, and functions.

The basic variable structure is a vector. Vectors can be created using the `<-` or `=` assignment operators (which assigns the evaluated expression on the right-hand side of the operator to the object name on the left-hand side).

```
> x <- c(5, 7, 9, 13, -4, 8)
> x = c(5, 7, 9, 13, -4, 8)  # equivalent
```

The above code creates a vector of length 6 using the `c()` function to concatenate scalars (2.2.10). The `=` operator must be used for the specification of options for functions. Other assignment operators exist, as well as the `assign()` function (see 4.1.4 or `help("<-")` for more information). The `rm()` command can be used to remove objects. The `exists()` function can be utilized to determine whether an object exists.

A.4.2 Indexing

Since vector operations are so common in R, it is important to be able to access (or index) elements within these vectors. Many different ways of indexing vectors are available. Here, we introduce several of these, using the above example. The command `x[2]` would return the second element of x (the scalar 7), and `x[c(2,4)]` would return the vector (7,13). The expressions `x[c(T,T,T,T,T,F)]`, `x[1:5]` and `x[-6]` would all return a vector consisting of the first five elements in x (the last specifies all elements except the 6th). Knowledge and basic comfort with these approaches to vector indexing are important to effective use of R, as they can help with computational efficiency.

Vectors are recycled if needed, for example, when comparing each of the elements of a vector to a scalar, as shown below.

```
> x>8
[1] FALSE FALSE  TRUE  TRUE FALSE FALSE
```

The above expression demonstrates the use of comparison operators (see `?Comparison`). Only the third and fourth elements of x are greater than 8. The function returns a logical value of either `TRUE` or `FALSE` (see `?Logic`).

A count of elements meeting the condition can be generated using the `sum()` function.

```
> sum(x>8)
[1] 2
```

The following commands create a vector of values greater than 8.

```
> largerthan8 = x[x>8]
> largerthan8
[1]  9 13
```

Here, the expression `x[x>8]` can be interpreted as "the elements of x for which x is greater than 8." This is a difficult construction for some new users. Examples of its application can be found in Sections 11.4.4.1 and 2.6.2.

Other comparison operators include `==` (equal), `>=` (greater than or equal), `<=` (less than or equal, and `!=` (not equal). Care needs to be taken in the comparison using `==` if noninteger values are present (see 3.2.5).

A.4.3   Operators

There are many operators defined in R to carry out a variety of tasks. Many of these were demonstrated in the sample session (assignment, arithmetic) and above examples (comparison). Arithmetic operations include `+`, `-`, `*`, `/`, `^` (exponentiation), `%%` (modulus), and `&/&` (integer division). More information about operators can be found using the help system (e.g., `?"+"`). Background information on other operators and precedence rules can be found using `help(Syntax)`.

R supports Boolean operations (OR, AND, NOT, and XOR) using the `|`, `||`, `&`, `!` operators and the `xor()` function. The `|` is an "or" operator that operates on each element of a vector, while the `||` is another "or" operator that stops evaluation the first time that the result is true (see `?Logic`).

A.4.4   Lists

Lists in R are generic objects that can contain other objects. List members can be named, or referenced using numeric indices (using the `[[` operator).

```
> newlist = list(x1="hello", x2=42, x3=TRUE)
> is.list(newlist)
[1] TRUE
> newlist
$x1
[1] "hello"

$x2
[1] 42

$x3
[1] TRUE
> newlist[[2]]
[1] 42
> newlist$x2
[1] 42
```

The `unlist()` function can be used to flatten (make a vector out of) the elements in a list (see also `relist()`).

```
> unlisted = unlist(newlist)
> unlisted
     x1      x2      x3
"hello"    "42"  "TRUE"
```

Note that unlisted objects are coerced (see 2.2.3) to a common type (in this case `character`).

A.4.5 Matrices

Matrices are rectangular objects with two dimensions (see 3.3). We can create a 2×3 matrix, display it, and test for its type.

```
> A = matrix(x, 2, 3)
> A
     [,1] [,2] [,3]
[1,]    5    9   -4
[2,]    7   13    8
> is.matrix(A)      # is A a matrix?
[1] TRUE
> is.vector(A)
[1] FALSE
> is.matrix(x)
[1] FALSE
```

Comments are supported within R (any input given after a **#** character is ignored).

Indexing for matrices is done in a similar fashion as for vectors, albeit with a second dimension (denoted by a comma).

```
> A[2,3]
[1] 8
> A[,1]
[1] 5 7
> A[1,]
[1]  5  9 -4
```

A.4.6 Dataframes

Analysis datasets are often stored in a dataframe, which is more general than a matrix. This rectangular object, similar to a dataset in other systems, can be thought of as a matrix with columns of vectors of different types (as opposed to a matrix, which consists of vectors of the same type). The functions **read.csv()** (see 1.1.4) and **read.table()** (see 1.1.2) return dataframe objects. A simple dataframe can be created using the **data.frame()** command. Access to sub-elements is achieved using the **$** operator, as shown below (see also **help(Extract)**). In addition, operations can be performed by column (e.g., calculation of sample statistics).

```
> y = rep(11, length(x))
> y
[1] 11 11 11 11 11 11
> ds = data.frame(x, y)
```

```
> ds
   x  y
1  5 11
```

```
2  7 11
3  9 11
4 13 11
5 -4 11
6  8 11
> ds$x[3]
[1] 9
```

We can check to see if an object is a dataframe with `is.data.frame()`. Note that the use of `data.frame()` differs from the use of `cbind()`, which yields a matrix object (unless `cbind()` is given dataframes as inputs).

```
> newmat = cbind(x, y)
> newmat
      x  y
[1,]  5 11
[2,]  7 11
[3,]  9 11
[4,] 13 11
[5,] -4 11
[6,]  8 11
```

```
> is.data.frame(newmat)
[1] FALSE
> is.matrix(newmat)
[1] TRUE
```

Dataframes can be thought of as the equivalent of datasets. They can be created from matrices using `as.data.frame()`, while matrices can be constructed from dataframes using `as.matrix()`.

Dataframes can be attached to the workspace using the `attach(ds)` command (see 2.1.1), though this is strongly discouraged [54]. After this command, individual columns in `ds` can be referenced directly by name (e.g., `x` instead of `ds$x`). Name conflicts are a common problem with `attach()` (see `conflicts()`, which reports on objects that exist with the same name in two or more places on the search path).

The `search()` function lists attached packages and objects. To avoid cluttering the name-space, the command `detach(ds)` should be used once a dataframe or package is no longer needed.

The `with()` and `within()` commands (see 2.1.1) can be used to simplify reference to an object within a dataframe without attaching.

The `sessionInfo()` function provides version information about R as well as details of loaded packages.

```
> sessionInfo()
R version 3.1.1 (2014-07-10)
Platform: x86_64-apple-darwin13.1.0 (64-bit)
locale:
[1] en_US.UTF-8/en_US.UTF-8/en_US.UTF-8/C/en_US.UTF-8/en_US.UTF-8
attached base packages:
[1] methods   stats    graphics  grDevices utils    datasets  base
```

The R.Version() function provides access to components of the version and platform status.

```
> R.Version()
$platform
[1] "x86_64-apple-darwin10.8.0"

$arch
[1] "x86_64"

$os
[1] "darwin10.8.0"

$system
[1] "x86_64, darwin10.8.0"

$status
[1] ""

$major
[1] "3"

$minor
[1] "1.1"

$year
[1] "2014"

$month
[1] "07"

$day
[1] "10"

$'svn rev'
[1] "66115"

$language
[1] "R"

$version.string
[1] "R version 3.1.1 (2014-07-10)"

$nickname
[1] "Sock it to Me"
```

Sometimes it is desirable to remove a package (A.6.1) from the workspace. For example, a package might define a function (4.2) with the same name as an existing function. Packages can be detached using the syntax detach(package:PKGNAME), where PKGNAME is the name of the package (see, for example, 7.10.5). Objects with the same name that appear in multiple places in the environment can be accessed using the location::objectname syntax. As an example, to access the `mean()` function from the `base` package, the user would specify `base::mean()` instead of `mean()`.

The names of all variables within a given dataset (or more generally for sub-objects within an object) are provided by the `names()` command. The names of all objects defined within an R session can be generated using the `objects()` and `ls()` commands, which return a vector of character strings. RStudio includes an `Environment` tab that lists all the objects in the current environment.

The `print()` and `summary()` functions can be used to display simple and more complex descriptions, respectively, of an object. Running `print(object)` at the command line is equivalent to just entering the name of the object, i.e., `object`.

A.4.7   Attributes and classes

Objects have a set of associated attributes (such as names of variables, dimensions, or classes) which can be displayed or sometimes changed. While a powerful concept, this can often be initially confusing. For example, we can find the dimension of the matrix defined earlier.

```
> attributes(A)
$dim
[1] 2 3
```

Other types of objects within R include lists (ordered objects that are not necessarily rectangular), regression models (objects of class `lm`), and formulae (e.g., `y ~ x1 + x2`). Examples of the use of formulas can be found in Sections 5.4.2 and 6.1.1. R supports object-oriented programming (see `help(UseMethod)`). As a result, objects within R have an associated "Class" attribute, which changes default behaviors for some operations on that object. Many functions have special capabilities when operating on a particular class. For example, when `summary()` is applied to an `lm` object, the `summary.lm()` function is called, while `summary.aov()` is called when an `aov` object is given as an argument. The `class()` function returns the classes to which an object belongs, while the `methods()` function displays all of the classes supported by a function (e.g., `methods(summary)`).

The `attributes()` command displays the attributes associated with an object, while the `typeof()` function provides information about the object (e.g., logical, integer, double, complex, character, and list). The `mode()` function displays the storage mode for an object.

A.4.8   Options

The `options()` function in R can be used to change various default behaviors, for example, the default number of digits to display in output (`options(digits=n)` where `n` is the preferred number). Defaults described in the book include `digits`, `show.signif.stars`, and `width`. The previous options are returned when `options()` is called (see 8.7.7), to allow them to be restored. The command `help(options)` lists all of the settable options.

A.5   Functions

A.5.1   Calling functions

Fundamental actions within R are carried out by calling functions (either built-in or user defined). Multiple arguments may be given, separated by commas. The function carries out operations using the provided arguments, then returns values (an object such as a vector or list) that are displayed (by default) or that can be saved by assignment to an object.

As an example, the `quantile()` function takes a vector and returns the minimum, 25th percentile, median, 75th percentile, and maximum, though if an optional vector of quantiles is given, those are calculated instead.

```
> vals = rnorm(1000) # generate 1000 standard normals
> quantile(vals)
     0%     25%     50%     75%    100%
-2.8290 -0.7017  0.0171  0.6577  3.8532
> quantile(vals, c(.025, .975))
 2.5% 97.5%
-2.03  1.89
```

Return values can be saved for later use.

```
> res = quantile(vals, c(.025, .975))
> res[1]
 2.5%
-2.03
```

Options are available for many functions. These are named arguments for the functions, and are generally added after the other arguments, also separated by commas. The documentation specifies the default action if named arguments (options) are not specified. For the `quantile()` function, there is a `type()` option that allows specification of one of nine algorithms for calculating quantiles. As an example, setting `type=3` specifies the "nearest even order statistic" option, which is the default for some systems (e.g., SAS).

```
> res = quantile(vals, probs=c(.025, .975), type=3)
```

Some functions allow a variable number of arguments. An example is the `paste()` function (see usage in 2.2.10). The calling sequence is described in the documentation as follows.

```
> paste(..., sep=" ", collapse=NULL)
```

To override the default behavior of a space being added between elements output by `paste()`, the user can specify a different value for `sep`.

A.5.2 The `apply` family of functions

Operations within R are most efficiently carried out using vector or list operations rather than looping. The `apply()` function can be used to perform many actions on an object. While somewhat subtle, the power of the vector language can be seen in this example. The `apply()` command is used to calculate column means or row means of the previously defined matrix in one fell swoop:

```
> A
     [,1] [,2] [,3]
[1,]    5    9   -4
[2,]    7   13    8
> apply(A, 2, mean)
[1]  6 11  2
> apply(A, 1, mean)
[1] 3.33 9.33
```

Option 2 specifies that the mean should be calculated for each column, while option 1 calculates the mean of each row. Here, we see some of the flexibility of the system, as functions in R (such as `mean()`) are also objects that can be passed as arguments to functions.

Other related functions include `lapply()`, which is helpful in avoiding loops when using lists; `sapply()` (see 2.1.2), `mapply()`, and `vapply()` to do the same for dataframes, matrices, and vectors, respectively; and `tapply()` (11.1.1) performs an action on subsets of an object. The `foreach` and `plyr` packages provide equivalent formulations for parallel execution (see also the `parallel` package).

A.5.3   Pipes and connections between functions

A recent addition to R is the pipe-forwarding mechanism (`%>%`) within the `magrittr` package. This is extremely useful when using the `dplyr`, `ggvis`, and `tidyr` packages, among others. Pipe forwarding is an alternative to nesting that yields code that can be read from top to bottom. A brief introduction can be found by running `vignette("magrittr")` (see 2.3.7 for an example).

Here we demonstrate an example that compares traditional (nested) `dplyr` function calls to the new pipe operator.

```
> library(dplyr)
> ds = read.csv("http://www.amherst.edu/~nhorton/r2/datasets/helpmiss.csv")
> summarise(group_by(select(filter(mutate(ds,
    sex=ifelse(female==1, "F", "M")), !is.na(pcs)), age, pcs, sex),
    sex), meanage=mean(age), meanpcs=mean(pcs),n=n())
Source: local data frame [2 x 4]

  sex meanage meanpcs   n
1   F    36.1    44.9 111
2   M    35.6    49.1 357
```

In this example, the output of the `mutate()` function is specified as the input to the `filter()` function, which prunes observations that are missing the `pcs` variable. The output from this function is sent to the `select()` function to create a subset of variables, and the results provided to the `group_by()` function, which collapse the dataset by gender. The `summarise()` function calculates the average age and PCS (physical component score) as well as the sample size.

This nested code is very difficult for humans to parse. An alternative would be to save the intermediate results from the nested functions.

```
> ds2 = mutate(ds, sex=ifelse(female==1, "F", "M"))
> ds3 = filter(ds2, !is.na(pcs))
> ds4 = select(ds3, age, pcs, sex)
```

```
> ds5 = group_by(ds4, sex)
> summarise(ds5, meanage=mean(age), meanpcs=mean(pcs),n=n())
Source: local data frame [2 x 4]

  sex meanage meanpcs   n
1   F    36.1    44.9 111
2   M    35.6    49.1 357
```

A disadvantage of this (somewhat clunky) approach is that it involves a lot of unnecessary copying. This may be particularly inefficient when processing large datasets.

The same operations are done in a different (and likely more readable) manner using the %>% operator.

```
> ds %>%
    mutate(sex=ifelse(female==1, "F", "M")) %>%
    filter(!is.na(pcs)) %>%
    select(age, pcs, sex) %>%
    group_by(sex) %>%
    summarise(meanage=mean(age), meanpcs=mean(pcs),n=n())
Source: local data frame [2 x 4]

  sex meanage meanpcs   n
1   F    36.1    44.9 111
2   M    35.6    49.1 357
```

Here, it is clear what each operation within the "pipe stream" is doing. It is straightforward to debug expressions in this manner by just leaving off the %>% at each line: this will only evaluate the set of functions called to that point and display the intermediate output.

A.6   Add-ons: packages

A.6.1   Introduction to packages

Additional functionality in R is added through packages, which consist of functions, datasets, examples, vignettes, and help files that can be downloaded from CRAN. The function install.packages() or the windowing interface under *Packages and Data* can be used to download and install packages. Alternatively, RStudio provides an easy-to-use **Packages** tab to install and load packages.

The library() function can be used to load a previously installed package (i.e., one that is included in the standard release of R or has been previously made available through use of the install.packages() function). As an example, to install and load Frank Harrell's Hmisc package, two commands are needed:

```
> install.packages("Hmisc")
> library(Hmisc)
```

Once a package has been installed, it can be loaded for use in a session of R by executing the function library(libraryname). If a package is not installed, running the library() command will yield an error. Here, we try to load the **Zelig** package (which had not yet been installed):

```
> library(Zelig)
Error in library(Zelig) : there is no package called 'Zelig'
```

To rectify the problem, we then install the package from CRAN.

```
> install.packages("Zelig")
trying URL 'ftp.osuosl.org/pub/cran/bin/mavericks/contrib/Zelig_4.2-1.tgz'
Content type 'application/x-gzip' length 3374792 bytes (3.2 Mb)
opened URL
==================================================
downloaded 3.2 Mb
The downloaded binary packages are in
/var/folders/2j/RtmpXPJ4o0/downloaded_packages
```

```
> library(Zelig)
ZELIG (Versions 4.2-1, built: 2013-09-12)
+-------------------------------------------------------------------+
|  Please refer to http://gking.harvard.edu/zelig for full          |
|  documentation or help.zelig() for help with commands and         |
|  models support by Zelig.                                         |
|  Zelig project citations:                                         |
|    Kosuke Imai, Gary King, and Olivia Lau.  (2009).               |
|    ''Zelig: Everyone's Statistical Software,''                    |
|    http://gking.harvard.edu/zelig                                 |
|  and                                                              |
|    Kosuke Imai, Gary King, and Olivia Lau. (2008).                |
|    ''Toward A Common Framework for Statistical Analysis           |
|    and Development,'' Journal of Computational and                |
|    Graphical Statistics, Vol. 17, No. 4 (December)                |
|    pp. 892-913.                                                   |
+-------------------------------------------------------------------+
Attaching package: 'Zelig'
```

Packages can be installed from other repositories (e.g., Omegahat or GitHub) by specifying the repository using the `repos=` option, or in the case of GitHub, using the `install_github()` function from the `devtools` package).

A user can test whether a package is available by running `require(packagename)`; this will load the library if it is installed, and generate a warning message if it is not (as opposed to `library()`, which will return an error, see 4.1.7). This is particularly useful in functions or reproducible analysis.

A.6.2 Packages and name conflicts

Different package authors may choose the same name for functions that exist within base R (or within other packages). This will cause the other function or object to be masked. This can sometimes lead to confusion, when the expected version of a function is not the one that is called. The `find()` function can be used to determine where in the environment (workspace) a given object can be found.

```
> find("mean")
[1] "package:base"
```

As an example where this might be useful, there are functions in the `base` and `Hmisc` packages called `units()`. The `find` command would display both (in the order which they would be accessed).

```
> library(Hmisc)
> find("units")
[1] "package:Hmisc" "package:base"
```

When the `Hmisc` package is loaded, the `units()` function from the `base` package is masked and would not be used by default. To specify that the version of the function from the `base` package should be used, prefix the function with the package name followed by two colons: `base::units()`. The `conflicts()` function reports on objects that exist with the same name in two or places on the search path.

A.6.3   Maintaining packages

The `update.packages()` function should be run periodically to ensure that packages are up to date (see `packageVersion()`). The `sessionInfo()` command displays the version of R that is running as well as information on all loaded packages.

The `packrat` package provides a comprehensive dependency system for R. This functionality can be extremely helpful to support reproducible analysis, as the exact set of packages used for an analysis can be identified and accessed in a project. Support for Packrat is built into RStudio.

As of October 2014, there were more than 5,900 packages available from CRAN. This represents a tremendous investment of time and code by many developers [44]. While each of these has met a minimal standard for inclusion, it is important to keep in mind that packages within R are created by individuals or small groups, and not endorsed by the R core group. As a result, they do not necessarily undergo the same level of testing and quality assurance that the core R system does.

A.6.4   CRAN task views

The *Task Views* on CRAN (`http://cran.r-project.org/web/views`) are a very useful resource for finding packages. These are listings of relevant packages within a particular application area (such as multivariate statistics, psychometrics, or survival analysis). Table A.2 displays the task views available as of October 2014.

A.6.5   Installed libraries and packages

Running the command `library(help="libraryname")` will display information about an installed package. Entries in the book that utilize packages include a line specifying how to access that library (e.g., `library(foreign)` in 1.1.6). As of October, 2014, the R distribution comes with the following packages:

base Base R functions

compiler R byte code compiler

datasets Base R datasets

Table A.2: CRAN task views

Bayesian	Bayesian inference
ChemPhys	Chemometrics and computational physics
Clinical Trials	Design, monitoring, and analysis of clinical trials
Cluster	Cluster analysis and finite mixture models
Differential Equations	Differential equations
Distributions	Probability distributions
Econometrics	Computational econometrics
Environmetrics	Analysis of ecological and environmental data
Experimental Design	Design and analysis of experiments
Finance	Empirical finance
Genetics	Statistical genetics
Graphics	Graphic displays, devices, and visualization
gR	Graphical models in R
High Performance Computing	High-performance and parallel computing
Machine Learning	Machine and statistical learning
Medical Imaging	Medical image analysis
MetaAnalysis	Meta-analysis
Multivariate	Multivariate statistics
Natural Language Processing	Natural language processing
Numerical Mathematics	Numerical mathematics
Official Statistics	Official statistics and survey methodology
Optimization	Optimization and mathematical programming
Pharmacokinetics	Analysis of pharmacokinetic data
Phylogenetics	Phylogenetics, especially comparative methods
Psychometrics	Psychometric models and methods
Reproducible Research	Reproducible research
Robust	Robust statistical methods
Social Sciences	Statistics for the social sciences
Spatial	Analysis of spatial data
Spatio Temporal	Handling and analyzing spatio-temporal data
Survival	Survival analysis
Time Series	Time series analysis
Web Technologies	Web technologies and service

graphics R functions for base graphics

grDevices Graphics devices for base and grid graphics

grid A rewrite of the graphics layout capabilities, plus some support for interaction

methods Formally defined methods and classes for R objects, plus other programming tools

parallel Support for parallel computation, including by forking and by sockets, and random-number generation

splines Regression spline functions and classes

stats R statistical functions

stats4 Statistical functions using S4 classes

tcltk Interface and language bindings to Tcl/Tk GUI elements

tools Tools for package development and administration

utils R utility functions

These packages are all available without having to run the `library()` command and are effectively part of R.

A.6.6 Packages referenced in this book

Other packages utilized in this book include:

biglm Bounded memory linear and generalized linear models [107]

boot Bootstrap functions [19] (recommended)

BRugs R interface to the OpenBUGS MCMC software [168]

car Companion to Applied Regression [45]

choroplethr Functions to simplify the creation of choropleths (thematic maps) in R [89]

chron Chronological objects [79]

circular Circular statistics [2]

coda Output analysis and diagnostics for Markov Chain Monte Carlo simulations [127]

coefplot Plots coefficients from fitted models [90]

coin Conditional inference procedures in a permutation test framework [75]

dispmod Dispersion models [155]

devtools Tools to make developing R code easier [193]

doBy Groupwise summary statistics, LSmeans, and general linear contrasts [66]

dplyr Plyr specialized for dataframes: faster and with remote datastores [194]

ellipse Functions for drawing ellipses and ellipse-like confidence regions [117]

elrm Exact logistic regression via MCMC [202]

epitools Epidemiology tools [7]

exactRankTests Exact distributions for rank and permutation tests [74]

factorplot Plot pairwise differences [8]

flexmix Flexible mixture modeling [96]

foreach Foreach looping construct for R [133]

foreign Read data stored by Minitab, S, SAS, SPSS, Stata, Systat, Weka, dBase [129] (recommended)

gam Generalized additive models [62]

gdata Various R programming tools for data manipulation [183]

gee Generalized estimation equation solver [21]

GenKern Functions for generating and manipulating binned kernel density estimates [105]

GGally Extension to ggplot2 [152]

ggmap A package for spatial visualization with Google Maps and OpenStreetMap [84]

ggplot2 An implementation of the grammar of graphics [188]

ggvis Implements a interactive grammar of graphics, taking the best parts of ggplot2, combining them with Shiny's reactive framework, and drawing web graphics using vega. [140]

gmodels Various R programming tools for model fitting [182]

greport Graphical reporting for clinical trials [83]

gridExtra Functions for grid graphics [9]

gtools Various R programming tools [184]

hexbin Hexagonal binning routines [22]

Hmisc Harrell miscellaneous [60]

Hotelling Hotelling's T-squared test and variants [29]

httr Tools for working with URLs and HTTP [191]

hwriter HTML writer: outputs R objects in HTML format [124]

irr Various coefficients of interrater reliability and agreement [46]

knitr A general-purpose package for dynamic report generation in R [200]

lars Least angle regression, LASSO, and forward stagewise [63]

lattice Lattice graphics [147] (recommended)

lawstat An R package for biostatistics, public policy, and law [49]

lme4 Linear mixed-effects models [12]

lmtest Testing linear regression models [203]

lpSolve Interface to Lp_solve v. 5.5 to solve linear/integer programs [16]

lubridate Makes dealing with dates a little easier [55]

magrittr A forward-pipe operator for R [10]

maps Draw geographical maps [15]

markdown Markdown rendering for R [5]

MASS Support functions and datasets for Venables and Ripley's MASS [179] (recommended)

Matching Multivariate and propensity score matching with balance optimization [157]

Matrix Sparse and dense matrix classes and methods [11] (recommended)

MCMCpack Markov Chain Monte Carlo (MCMC) package [110]

memisc Tools for survey data, graphics, programming, statistics, and simulation [34]

mice Multivariate imputation by chained equations [178]

mitools Tools for multiple imputation of missing data [108]

mix Estimation/multiple imputation for mixed categorical and continuous data [150]

moments Moments, cumulants, skewness, kurtosis, and related tests [88]

mosaic Project MOSAIC statistics and mathematics teaching utilities [128]

MplusAutomation Automating Mplus model estimation and interpretation [58]

muhaz Hazard function estimation in survival analysis [64]

multcomp Simultaneous inference in general parametric models [73]

multilevel Multilevel functions [17]

nlme Linear and nonlinear mixed-effects models [125] (recommended)

nnet Feed-forward neural networks and multinomial log-linear models [179] (recommended)

nortest Tests for normality [56]

packrat A dependency management system for projects and their R package dependencies [175]

partykit A toolkit for recursive partytioning [76]

plotrix Various plotting functions [97]

plyr Tools for splitting, applying and combining data [190]

poLCA Polytomous variable latent class analysis [102]

prettyR Pretty descriptive stats [98]

pscl Political science computational laboratory, Stanford University [78]

pwr Basic functions for power analysis [23]

QuantPsyc Quantitative psychology tools [42]

quantreg Quantile regression [87]

R2jags A package for running jags from R [163]

R2WinBUGS Running WinBUGS and OpenBUGS from R [162]

randomLCA Random effects latent class analysis [14]

RCurl General network (HTTP/FTP) client interface for R [92]

reshape Flexibly reshape data [187]

rjags Bayesian graphical models using MCMC [126]

RMongo MongoDB client for R [24]

rms Regression modeling strategies [61]

RMySQL R interface to the MySQL database [80]

ROCR Visualizing the performance of scoring classifiers [161]

RODBC An ODBC database interface [134]

rpart Recursive partitioning [167] (recommended)

RSQLite SQLite interface for R [81]

rtf Rich text format (RTF) output [151]

runjags Interface utilities, parallel computing methods, and additional distributions for MCMC models in JAGS [31]

sas7bdat SAS database reader [159]

scatterplot3d 3D scatter plot [101]

sciplot Scientific graphing functions for factorial designs [115]

simPH Tools for simulating and plotting quantities of interest estimated from Cox proportional hazards models [48]

shiny Shiny makes it incredibly easy to build interactive web applications with R. Automatic "reactive" binding between inputs and outputs and extensive prebuilt widgets make it possible to build beautiful, responsive, and powerful applications with minimal effort. [141]

sqldf Perform SQL selects on R dataframes [57]

survey Analysis of complex survey samples [106]

survival Survival analysis [166] (recommended)

swirl Learn R, in R [20]

tidyr Easily tidy data with spread and gather functions [192].

tm A framework for text mining applications within R [38]

tmvtnorm Truncated multivariate normal and Student t distribution [195]

vcd Visualizing categorical data [113]

VGAM Vector generalized linear and additive models [201]

vioplot Violin plot [1]

WriteXLS Cross-platform Perl-based R function to create Excel spreadsheets [153]

XML Tools for parsing and generating XML [91]

xtable Export tables to LaTeX or HTML [30]

Zelig Everyone's statistical software [123]

These must be downloaded, installed, and loaded prior to use (see `install.packages()`, `require()`, and `library()`), though the recommended packages are included in most distributions of R. To facilitate the process of loading the other packages, we have created a script file to load these in one step (see A.2.1).

A.6.7   Datasets available with R

A number of datasets are available within the `datasets` package. The `data()` function lists these, while the optional `package` option can be used to regenerate datasets from within a specific package.

A.7   Support and bugs

Since R is a free software project written by volunteers, there are no paid support options available directly from the R Foundation. A number of groups provide commercial support for R and related systems, including Revolution Analytics and RStudio. In addition, extensive resources are available to help users.

In addition to the manuals, publications, FAQs, newsletter, task views, and books listed on the `www.r-project.org` web page, there are a number of mailing lists that exist to help answer questions. Because of the volume of postings, it is important to carefully read the posting guide at `http://www.r-project.org/posting-guide.html` prior to submitting a question. These guidelines are intended to help leverage the value of the list, to avoid embarrassment, and to optimize the allocation of limited resources to technical issues.

As in any general-purpose statistical software package, some bugs exist. More information about the process of determining whether and how to report a problem can be found using `help(bug.report)` (please also review the R FAQ).

Appendix B

The HELP study dataset

B.1 Background on the HELP study

Data from the HELP (Health Evaluation and Linkage to Primary Care) study are used to illustrate many of the entries. The HELP study was a clinical trial for adult inpatients recruited from a detoxification unit. Patients with no primary care physician were randomized to receive a multidisciplinary assessment and a brief motivational intervention or usual care, with the goal of linking them to primary medical care. Funding for the HELP study was provided by the National Institute on Alcohol Abuse and Alcoholism (R01-AA10870, Samet PI) and the National Institute on Drug Abuse (R01-DA10019, Samet PI).

Eligible subjects were adults, who spoke Spanish or English, reported alcohol, heroin, or cocaine as their first or second drug of choice, and either resided in proximity to the primary care clinic to which they would be referred, or were homeless. Patients with established primary care relationships they planned to continue, significant dementia, specific plans to leave the Boston area that would prevent research participation, failure to provide contact information for tracking purposes, or pregnancy were excluded.

Subjects were interviewed at baseline during their detoxification stay, and follow-up interviews were undertaken every 6 months for 2 years. A variety of continuous, count, discrete, and survival time predictors and outcomes were collected at each of these five occasions.

The details of the randomized trial along with the results from a series of additional analyses have been published [145, 132, 72, 100, 85, 144, 143, 158, 93, 198].

B.2 Roadmap to analyses of the HELP dataset

Table B.1 summarizes the analyses illustrated using the HELP dataset. These analyses are intended to help illustrate the methods described in the book. Interested readers are encouraged to review the published data from the HELP study for substantive analyses.

Table B.1: Analyses undertaken using the HELP dataset

Description	Section (page)
Data input and output	2.6.1 (p. 25)
Summarize data contents	2.6.1 (p. 25)
Data display	2.6.1 (p. 26)
Derived variables and data manipulation	2.6.3 (p. 27)

Description	Section (page)
Sorting and subsetting	2.6.4 (p. 31)
Summary statistics	5.7.1 (p. 59)
Exploratory data analysis	5.7.1 (p. 59)
Bivariate relationship	5.7.2 (p. 60)
Contingency tables	5.7.3 (p. 61)
Two-sample tests	5.7.4 (p. 64)
Survival analysis (logrank test)	5.7.5 (p. 66)
Scatterplot with smooth fit	6.6.1 (p. 76)
Linear regression with interaction	6.6.2 (p. 77)
Regression coefficient plot	6.6.3 (p. 81)
Regression diagnostics	6.6.4 (p. 81)
Fitting stratified regression models	6.6.5 (p. 83)
Two-way analysis of variance (ANOVA)	6.6.6 (p. 84)
Multiple comparisons	6.6.7 (p. 87)
Contrasts	6.6.8 (p. 88)
Logistic regression	7.10.1 (p. 104)
Poisson regression	7.10.2 (p. 105)
Zero-inflated Poisson regression	7.10.3 (p. 106)
Negative binomial regression	7.10.4 (p. 107)
Quantile regression	7.10.5 (p. 107)
Ordinal logit	7.10.6 (p. 108)
Multinomial logit	7.10.7 (p. 108)
Generalized additive model	7.10.8 (p. 109)
Reshaping datasets	7.10.9 (p. 110)
General linear model for correlated data	7.10.10 (p. 112)
Random effects model	7.10.11 (p. 113)
Generalized estimating equations model	7.10.12 (p. 115)
Generalized linear mixed model	7.10.13 (p. 116)
Proportional hazards regression model	7.10.14 (p. 117)
Cronbach α	7.10.15 (p. 117)
Factor analysis	7.10.16 (p. 118)
Recursive partitioning	7.10.17 (p. 119)
Linear discriminant analysis	7.10.18 (p. 120)
Hierarchical clustering	7.10.19 (p. 121)
Scatterplot with multiple y axes	8.7.1 (p. 134)
Conditioning plot	8.7.2 (p. 135)
Scatterplot with marginal histogram	8.7.3 (p. 135)
Kaplan–Meier plot	8.7.4 (p. 137)
ROC curve	8.7.5 (p. 138)
Pairs plot	8.7.6 (p. 138)
Visualize correlation matrix	8.7.7 (p. 141)
By group processing	11.1.2 (p. 168)
Bayesian regression	11.4.1 (p. 173)
Propensity score modeling	11.4.2 (p. 177)
Multiple imputation	11.4.4.2 (p. 183)
Interactive visualization	12.6 (p. 203)

B.3 Detailed description of the dataset

The Institutional Review Board of Boston University Medical Center approved all aspects of the study, including the creation of the de-identified dataset. Additional privacy protection was secured by the issuance of a Certificate of Confidentiality by the Department of Health and Human Services.

A de-identified dataset containing the variables utilized in the end-of-chapter examples is available for download at the book web site: `http://www.amherst.edu/~nhorton/r2/datasets/help.csv`.

Variables included in the HELP dataset are described in Table B.2. A full copy of the study instruments can be found at `http://www.amherst.edu/~nhorton/help`.

Table B.2: Annotated description of variables in the HELP dataset

VARIABLE	DESCRIPTION	VALUES	NOTE
a15a	Number of nights in overnight shelter in past 6 months	0–180	See also `homeless`
a15b	Number of nights on the street in past 6 months	0–180	See also `homeless`
age	Age at baseline (in years)	19–60	
anysubstatus	Use of any substance post-detox	0=no, 1=yes	See also `daysanysub`
cesd*	Center for Epidemiologic Studies Depression scale	0–60	Higher scores indicate more depressive symptoms; see also `f1a`–`f1t`.
d1	How many times hospitalized for medical problems (lifetime)	0–100	
daysanysub	Time (in days) to first use of any substance post-detox	0–268	See also `anysubstatus`
daysdrink	Time (in days) to first alcoholic drink post-detox	0–270	See also `drinkstatus`
dayslink	Time (in days) to linkage to primary care	0–456	See also `linkstatus`
drinkstatus	Use of alcohol post-detox	0=no, 1=yes	See also `daysdrink`
drugrisk*	Risk-Assessment Battery (RAB) drug risk score	0–21	Higher scores indicate riskier behavior; see also `sexrisk`.
e2b*	Number of times in past 6 months entered a detox program	1–21	
f1a	I was bothered by things that usually don't bother me.	0–3[#]	
f1b	I did not feel like eating; my appetite was poor.	0–3[#]	
f1c	I felt that I could not shake off the blues even with help from my family or friends.	0–3[#]	
f1d	I felt that I was just as good as other people.	0–3[#]	

VARIABLE	DESCRIPTION	VALUES	NOTE
f1e	I had trouble keeping my mind on what I was doing.	0–3[#]	
f1f	I felt depressed.	0–3[#]	
f1g	I felt that everything I did was an effort.	0–3[#]	
f1h	I felt hopeful about the future.	0–3[#]	
f1i	I thought my life had been a failure.	0–3[#]	
f1j	I felt fearful.	0–3[#]	
f1k	My sleep was restless.	0–3[#]	
f1l	I was happy.	0–3[#]	
f1m	I talked less than usual.	0–3[#]	
f1n	I felt lonely.	0–3[#]	
f1o	People were unfriendly.	0–3[#]	
f1p	I enjoyed life.	0–3[#]	
f1q	I had crying spells.	0–3[#]	
f1r	I felt sad.	0–3[#]	
f1s	I felt that people dislike me.	0–3[#]	
f1t	I could not get going.	0–3[#]	
female	Gender of respondent	0=male, 1=female	
g1b*	Experienced serious thoughts of suicide (last 30 days)	0=no, 1=yes	
homeless*	1 or more nights on the street or shelter in past 6 months	0=no, 1=yes	See also a15a and a15b
i1*	Average number of drinks (standard units) consumed per day (in the past 30 days)	0–142	See also i2
i2	Maximum number of drinks (standard units) consumed per day (in the past 30 days)	0–184	See also i1
id	Random subject identifier	1–470	
indtot*	Inventory of Drug Use Consequences (InDUC) total score	4–45	
linkstatus	Post-detox linkage to primary care	0=no, 1=yes	See also dayslink
mcs*	SF-36 Mental Component Score	7-62	Higher scores indicate better functioning; see also pcs.
pcrec*	Number of primary care visits in past 6 months	0–2	See also linkstatus, not observed at baseline.
pcs*	SF-36 Physical Component Score	14-75	Higher scores indicate better functioning; see also mcs.
pss_fr	Perceived social supports (friends)	0–14	

VARIABLE	DESCRIPTION	VALUES	NOTE
satreat	Any BSAS substance abuse treatment at baseline	0=no, 1=yes	
sexrisk*	Risk-Assessment Battery (RAB) sex risk score	0–21	Higher scores indicate riskier behavior; see also drugrisk.
substance	Primary substance of abuse	alcohol, cocaine, or heroin	
treat	Randomization group	0=usual care, 1=HELP clinic	

Notes: Observed range is provided (at baseline) for continuous variables.

* Denotes variables measured at baseline and follow-up (e.g., cesd is baseline measure, cesd1 is measured at 6 months, and cesd4 is measured at 24 months).

For each of the 20 items in HELP Section F1 (CESD), respondents were asked to indicate how often they behaved this way during the past week (0 = rarely or none of the time, less than 1 day; 1 = some or a little of the time, 1–2 days; 2 = occasionally or a moderate amount of time, 3–4 days; or 3 = most or all of the time, 5–7 days); items f1d, f1h, f1l, and f1p were reverse coded.

Appendix C

References

[1] D. Adler. *vioplot: Violin plot*, 2005. R package version 0.2.

[2] C. Agostinelli and U. Lund. *R package circular: Circular Statistics (version 0.4-7)*, 2013.

[3] A. Agresti. *Categorical Data Analysis*. John Wiley & Sons, Hoboken, NJ, 2002.

[4] J. Albert. *Bayesian Computation with R*. Springer, New York, 2008.

[5] J. J. Allaire, J. Horner, V. Marti, and N. Porte. *markdown: Markdown Rendering for R*, 2014. R package version 0.7.4.

[6] D. G. Altman and J.M. Bland. Measurement in medicine: the analysis of method comparison studies. *The Statistician*, 32:307–317, 1983.

[7] T. J. Aragon. *epitools: Epidemiology Tools*, 2012. R package version 0.5-7.

[8] D. Armstrong. *factorplot: factorplot*, 2014. R package version 1.1-1.

[9] B. Auguie. *gridExtra: Functions in Grid Graphics*, 2012. R package version 0.9.1.

[10] S. B. Bache and H. Wickham. *magrittr: A Forward-Pipe Operator for R*, 2014. R package version 1.0.1.

[11] D. Bates and M. Maechler. *Matrix: Sparse and Dense Matrix Classes and Methods*, 2014. R package version 1.1-4.

[12] D. Bates, M. Maechler, B. Bolker, and S. Walker. *lme4: Linear Mixed-Effects Models Using Eigen and S4*, 2014. R package version 1.1-7.

[13] B. Baumer, M. Çetinkaya Rundel, A. Bray, L. Loi, and N. J. Horton. R markdown: Integrating a reproducible analysis tool into introductory statistics. *Technology Innovations in Statistics Education*, 8(1), 2014.

[14] K. Beath. *randomLCA: Random Effects Latent Class Analysis*, 2014. R package version 0.9-0.

[15] R. A. Becker, A. R. Wilks, R. Brownrigg, and T. P. Minka. *maps: Draw Geographical Maps*, 2014. R package version 2.3-9.

[16] M. Berkelaar. *lpSolve: Interface to Lp_solve v. 5.5 to Solve Linear/Integer Programs*, 2014. R package version 5.6.10.

[17] P. Bliese. *multilevel: Multilevel Functions*, 2013. R package version 2.5.

[18] T. S. Breusch and A. R. Pagan. A simple test for heteroscedasticity and random coefficient variation. *Econometrica*, 47, 1979.

[19] A. Canty and B. Ripley. *boot: Bootstrap R (S-Plus) Functions*, 2014. R package version 1.3-13.

[20] N. Carchedi, B. Bauer, G. Grdina, and S. Kross. *swirl: Learn R, in R*, 2014. R package version 2.2.16.

[21] V. J. Carey. *gee: Generalized Estimation Equation Solver*, 2012. R package version 4.13-18.

[22] D. Carr, N. Lewin-Koh, and M. Maechler. *hexbin: Hexagonal Binning Routines*, 2014. R package version 1.27.0.

[23] S. Champely. *pwr: Basic Functions for Power Analysis*, 2012. R package version 1.1.1.

[24] T. Chheng. *RMongo: MongoDB Client for R*, 2013. R package version 0.0.25.

[25] D. Collett. *Modelling Binary Data*. Chapman & Hall, London, 1991.

[26] D. Collett. *Modeling Survival Data in Medical Research (second edition)*. CRC Press, Boca Raton, FL, 2003.

[27] L. M. Collins, J. L. Schafer, and C.-M. Kam. A comparison of inclusive and restrictive strategies in modern missing data procedures. *Psychological Methods*, 6(4):330–351, 2001.

[28] R. D. Cook. *Residuals and Influence in Regression*. Chapman & Hall, London, 1982.

[29] J. M. Curran. *Hotelling's T-squared Test and Variants*, 2013. R package version 1.0-2.

[30] D. B. Dahl. *xtable: Export Tables to LaTeX or HTML*, 2014. R package version 1.7-4.

[31] M. J. Denwood. runjags: An R package providing interface utilities, parallel computing methods and additional distributions for MCMC models in JAGS. *Journal of Statistical Software*, In Review.

[32] A. J. Dobson and A. Barnett. *An Introduction to Generalized Linear Models (third edition)*. CRC Press, Boca Raton, FL, 2008.

[33] B. Efron and R. J. Tibshirani. *An Introduction to the Bootstrap*. Chapman & Hall, London, 1993.

[34] M. Elff. *memisc: Tools for Management of Survey Data, Graphics, Programming, Statistics, and Simulation*, 2013. R package version 0.96-9.

[35] M. J. Evans and J. S. Rosenthal. *Probability and Statistics: The Science of Uncertainty*. W H Freeman and Company, New York, 2004.

[36] J. J. Faraway. *Linear Models with R*. CRC Press, Boca Raton, FL, 2004.

[37] J. J. Faraway. *Extending the Linear Model with R: Generalized Linear, Mixed Effects and Nonparametric Regression Models*. CRC Press, Boca Raton, FL, 2005.

[38] I. Feinere, K. Hornik, and D. Meyer. Text mining infrastructure in R. *Journal of Statistical Software*, 25(5):1–54, 2008.

[39] N. I. Fisher. *Statistical Analysis of Circular Data*. Cambridge University Press, New York, 1996.

[40] G. M. Fitzmaurice, N. M. Laird, and J. H. Ware. *Applied Longitudinal Analysis*. John Wiley & Sons, Hoboken, NJ, 2004.

[41] T. R. Fleming and D. P. Harrington. *Counting Processes and Survival Analysis*. John Wiley & Sons, Hoboken, NJ, 1991.

[42] T. D. Fletcher. *QuantPsyc: Quantitative Psychology Tools*, 2012. R package version 1.5.

[43] J. Fox. The R Commander: a basic graphical user interface to R. *Journal of Statistical Software*, 14(9), 2005.

[44] J. Fox. Aspects of the social organization and trajectory of the R Project. *The R Journal*, 1(2):5–13, December 2009.

[45] John Fox and Sanford Weisberg. *An R Companion to Applied Regression (second edition)*. Sage, Thousand Oaks, CA, 2011.

[46] M. Gamer, J. Lemon, I. Fellows, and P. Singh. *irr: Various Coefficients of Interrater Reliability and Agreement*, 2012. R package version 0.84.

[47] C. Gandrud. *Reproducible Research with R and RStudio*. CRC Press, Boca Raton, FL, 2014.

[48] C. Gandrud. *simPH: Tools for Simulating and Plotting Quantities of Interest Estimated from Cox Proportional Hazards Models*, 2014. R package version 1.2.3.

[49] J. L. Gastwirth, Y. R. Gel, W. L. Wallace Hui, V. Lyubchich, W. Miao, and K. Noguchi. *lawstat: An R Package for Biostatistics, Public Policy, and Law*, 2013. R package version 2.4.1.

[50] A. Gelman, J. B. Carlin, H. S. Stern, and D. B. Rubin. *Bayesian Data Analysis (second edition)*. Chapman & Hall, London, 2004.

[51] A. Gelman, C. Pasarica, and R. Dodhia. Let's practice what we preach: turning tables into graphs. *The American Statistician*, 56:121–130, 2002.

[52] R. Gentleman and D. Temple Lang. Statistical analyses and reproducible research. *Journal of Computational and Graphical Statistics*, 16(1):1–23, 2007.

[53] L. Gonick. *Cartoon Guide to Statistics*. HarperPerennial, New York, 1993.

[54] Google. R style guide. http://google-styleguide.googlecode.com/svn/trunk/Rguide.xml, date accessed 10/29/2013, 2013.

[55] G. Grolemund and H. Wickham. Dates and times made easy with lubridate. *Journal of Statistical Software*, 40(3):1–25, 2011.

[56] J. Gross and U. Ligges. *nortest: Tests for Normality*, 2012. R package version 1.0-2.

[57] G. Grothendieck. *sqldf: Perform SQL Selects on R Data Frames*, 2014. R package version 0.4-7.1.

[58] M. Hallquist and J. Wiley. *MplusAutomation: Automating Mplus Model Estimation and Interpretation*, 2013. R package version 0.6-2.

[59] J. W. Hardin and J. M. Hilbe. *Generalized Estimating Equations*. CRC Press, Boca Raton, FL, 2002.

[60] F. E. Harrell. *Hmisc: Harrell Miscellaneous*, 2014. R package version 3.14-5.

[61] F. E. Harrell. *rms: Regression Modeling Strategies*, 2014. R package version 4.2-1.

[62] T. Hastie. *gam: Generalized Additive Models*, 2014. R package version 1.09.1.

[63] T. Hastie and B. Efron. *lars: Least Angle Regression, Lasso and Forward Stagewise*, 2013. R package version 1.2.

[64] K. Hess and R. Gentleman. *muhaz: Hazard Function Estimation in Survival Analysis*, 2014. R package version 1.2.6.

[65] T. C. Hesterberg, D. S. Moore, S. Monaghan, A. Clipson, and R. Epstein. *Bootstrap Methods and Permutation Tests*. W.C. Freeman, 2005.

[66] S. Højsgaard and U. Halekoh. *doBy: Groupwise Summary Statistics, LSmeans, General Linear Contrasts, Various Utilities*, 2014. R package version 4.5-11.

[67] N. J. Horton. I hear, I forget. I do, I understand: A modified Moore-method mathematical statistics course. *The American Statistician*, 67(3):219–228, 2013.

[68] N. J. Horton, E. R. Brown, and L. Qian. Use of R as a toolbox for mathematical statistics exploration. *The American Statistician*, 58(4):343–357, 2004.

[69] N. J. Horton, E. Kim, and R. Saitz. A cautionary note regarding count models of alcohol consumption in randomized controlled trials. *BMC Medical Research Methodology*, 7(9), 2007.

[70] N. J. Horton and K. P. Kleinman. Much ado about nothing: A comparison of missing data methods and software to fit incomplete data regression models. *The American Statistician*, 61:79–90, 2007.

[71] N. J. Horton and S. R. Lipsitz. Multiple imputation in practice: comparison of software packages for regression models with missing variables. *The American Statistician*, 55(3):244–254, 2001.

[72] N. J. Horton, R. Saitz, N. M. Laird, and J. H. Samet. A method for modeling utilization data from multiple sources: application in a study of linkage to primary care. *Health Services and Outcomes Research Methodology*, 3:211–223, 2002.

[73] T. Hothorn, F. Bretz, and P. Westfall. Simultaneous inference in general parametric models. *Biometrical Journal*, 50(3):346–363, 2008.

[74] T. Hothorn and K. Hornik. *exactRankTests: Exact Distributions for Rank and Permutation Tests*, 2013. R package version 0.8-27.

[75] T. Hothorn, K. Hornik, M. A. van de Wiel, and A. Zeileis. Implementing a class of permutation tests: The coin package. *Journal of Statistical Software*, 28(8):1–23, 2008.

[76] T. Hothorn and A. Zeileis. *partykit: A Toolkit for Recursive Partytioning*, 2014. R package version 0.8-2.

[77] R. Ihaka and R. Gentleman. R: A language for data analysis and graphics. *Journal of Computational and Graphical Statistics*, 5(3):299–314, 1996.

[78] S. Jackman. *pscl: Classes and Methods for R Developed in the Political Science Computational Laboratory, Stanford University*, 2014. R package version 1.4.6.

[79] D. James and K. Hornik. *chron: Chronological Objects Which Can Handle Dates and Times*, 2014. R package version 2.3-45. S original by David James, R port by Kurt Hornik.

[80] D. A. James and S. DebRoy. *RMySQL: R Interface to the MySQL Database*, 2012. R package version 0.9-3.

[81] D. A. James and S. Falcon. *RSQLite: SQLite Interface for R*, 2013. R package version 0.11.4.

[82] S. R. Jammalamadaka and A. Sengupta. *Topics in Circular Statistics*. World Scientific, River Edge, NJ, 2001.

[83] F. E. Harrell Jr. *greport: Graphical Reporting for Clinical Trials*, 2014. R package version 0.5-1.

[84] D. Kahle and H. Wickham. *ggmap: A package for spatial visualization with Google Maps and OpenStreetMap*, 2013. R package version 2.3.

[85] S. G. Kertesz, N. J. Horton, P. D. Friedmann, R. Saitz, and J. H. Samet. Slowing the revolving door: stabilization programs reduce homeless persons substance use after detoxification. *Journal of Substance Abuse Treatment*, 24:197–207, 2003.

[86] D. Knuth. Literate programming. *CSLI Lecture Notes*, 27, 1992.

[87] R. Koenker. *quantreg: Quantile Regression*, 2013. R package version 5.05.

[88] L. Komsta and F. Novomestky. *moments: Moments, Cumulants, Skewness, Kurtosis and Related Tests*, 2012. R package version 0.13.

[89] A. Lamstein and B.P. Johnson. *Functions to Simplify the Creation of Choropleths (Thematic Maps) in R*, 2014. R package version 1.7.0.

[90] J. P. Lander. *coefplot: Plots Coefficients from Fitted Models*, 2013. R package version 1.2.0.

[91] D. Temple Lang. *XML: Tools for Parsing and Generating XML within R and S-Plus*, 2013. R package version 3.98-1.1.

[92] D. Temple Lang. *RCurl: General Network (HTTP/FTP/...) Client Interface for R*, 2014. R package version 1.95-4.3.

[93] M. J. Larson, R. Saitz, N. J. Horton, C. Lloyd-Travaglini, and J. H. Samet. Emergency department and hospital utilization among alcohol and drug-dependent detoxification patients without primary medical care. *American Journal of Drug and Alcohol Abuse*, 32:435–452, 2006.

[94] M. Lavine. *Introduction to Statistical Thought*. http://www.math.umass.edu/~lavine/Book/book.html, 2005.

[95] F. Leisch. Sweave: Dynamic generation of statistical reports using literate data analysis. In Wolfgang Härdle and Bernd Rönz, editors, *Compstat 2002 — Proceedings in Computational Statistics*, pages 575–580. Physica Verlag, Heidelberg, 2002.

[96] F. Leisch. FlexMix: A general framework for finite mixture models and latent class regression in R. *Journal of Statistical Software*, 11(8):1–18, 2004.

[97] J. Lemon. Plotrix: a package in the red light district of R. *R-News*, 6(4):8–12, 2006.

[98] J. Lemon and P. Grosjean. *prettyR: Pretty Descriptive Stats*, 2014. R package version 2.0-8.

[99] K.-Y. Liang and S. L. Zeger. Longitudinal data analysis using generalized linear models. *Biometrika*, 73:13–22, 1986.

[100] J. Liebschutz, J. B. Savetsky, R. Saitz, N. J. Horton, C. Lloyd-Travaglini, and J. H. Samet. The relationship between sexual and physical abuse and substance abuse consequences. *Journal of Substance Abuse Treatment*, 22(3):121–128, 2002.

[101] U. Ligges and M. Mächler. Scatterplot3d: an R package for visualizing multivariate data. *Journal of Statistical Software*, 8(11):1–20, 2003.

[102] D. A. Linzer and J. B. Lewis. poLCA: an R package for polytomous variable latent class analysis. *Journal of Statistical Software*, 42(10):1–29, 2011.

[103] S. R. Lipsitz, N. M. Laird, and D. P. Harrington. Maximum likelihood regression methods for paired binary data. *Statistics in Medicine*, 9:1517–1525, 1990.

[104] R. H. Lock, P. F. Lock, K. L. Lock, E. F. Lock, and D. F. Lock. *Statistics: Unlocking the Power of Data*. John Wiley & Sons, Hoboken, NJ, 2013.

[105] D. Lucy and R. Aykroyd. *GenKern: Functions for Generating and Manipulating Binned Kernel Density Estimates*, 2013. R package version 1.2-60.

[106] T. Lumley. Analysis of complex survey samples. *Journal of Statistical Software*, 9(1):1–19, 2004.

[107] T. Lumley. *biglm: Bounded Memory Linear and Generalized Linear Models*, 2013. R package version 0.9-1.

[108] T. Lumley. *mitools: Tools for Multiple Imputation of Missing Data*, 2014. R package version 2.3.

[109] B. F. J. Manly. *Multivariate Statistical Methods: A Primer (third edition)*. CRC Press, Boca Raton, FL, 2004.

[110] A. D. Martin, K. M. Quinn, and J. H. Park. MCMCpack: Markov Chain Monte Carlo in R. *Journal of Statistical Software*, 42(9):22, 2011.

[111] P. McCullagh and J. A. Nelder. *Generalized Linear Models*. Chapman & Hall, London, 1989.

[112] N. Metropolis, A.W. Rosenbluth, A.H. Teller, and E. Teller. Equations of state calculations by fast computing machines. *Journal of Chemical Physics*, 21(6):1087–1092, 1953.

[113] D. Meyer, A Zeileis, and Kurt Hornik. The strucplot framework: visualizing multi-way contingency tables with vcd. *Journal of Statistical Software*, 17(3):1–48, 2006.

[114] J. D. Mills. Using computer simulation methods to teach statistics: a review of the literature. *Journal of Statistics Education*, 10(1), 2002.

[115] M. Morales. *sciplot: Scientific Graphing Functions for Factorial Designs*, 2012. R package version 1.1-0.

[116] F. Mosteller. *Fifty Challenging Problems in Probability with Solutions*. Dover Publications, 1987.

[117] D. Murdoch and E. D. Chow. *ellipse: Functions for Drawing Ellipses and Ellipse-Like Confidence Regions*, 2013. R package version 0.3-8.

[118] P. Murrell. *R Graphics*. Chapman & Hall, London, 2005.

[119] P. Murrell. *Introduction to Data Technologies*. Chapman & Hall, London, 2009.

[120] N. J. D. Nagelkerke. A note on a general definition of the coefficient of determination. *Biometrika*, 78(3):691–692, 1991.

[121] National Institutes of Alcohol Abuse and Alcoholism, Bethesda, MD. *Helping Patients Who Drink Too Much*, 2005.

[122] D. Nolan and D. Temple Lang. *XML and Web Technologies for Data Sciences with R*. Springer, New York, 2014.

[123] M. Owen, K. Imai, G. King, and O. Lau. *Zelig: Everyone's Statistical Software*, 2013. R package version 4.2-1.

[124] G. Pau. *hwriter: HTML Writer: Outputs R Objects in HTML Format*, 2014. R package version 1.3.2.

[125] J. Pinheiro, D. Bates, S. DebRoy, and D. Sarkar. *nlme: Linear and Nonlinear Mixed Effects Models*, 2014. R package version 3.1-117.

[126] M. Plummer. *rjags: Bayesian Graphical Models Using MCMC*, 2014. R package version 3-13.

[127] M. Plummer, N. Best, K. Cowles, and K. Vines. Coda: convergence diagnosis and output analysis for MCMC. *R News*, 6(1):7–11, 2006.

[128] R. Pruim, D. Kaplan, and N. J. Horton. *mosaic: Project MOSAIC (mosaic-web.org) Statistics and Mathematics Teaching Utilities*, 2014. R package version 0.9-1-3.

[129] R Core Team. *foreign: Read Data Stored by Minitab, S, SAS, SPSS, Stata, Systat, Weka, dBase, ...*, 2014. R package version 0.8-61.

[130] R Development Core Team. *R: A Language and Environment for Statistical Computing*. R Foundation for Statistical Computing, Vienna, 2013.

[131] T. E. Raghunathan, J. M. Lepkowski, J. van Hoewyk, and P. Solenberger. A multivariate technique for multiply imputing missing values using a sequence of regression models. *Survey Methodology*, 27(1):85–95, 2001.

[132] V. W. Rees, R. Saitz, N. J. Horton, and J. H. Samet. Association of alcohol consumption with HIV sex and drug risk behaviors among drug users. *Journal of Substance Abuse Treatment*, 21(3):129–134, 2001.

[133] Revolution Analytics and S. Weston. *foreach: Foreach Looping Construct for R*, 2014. R package version 1.4.2.

[134] B. Ripley and M. Lapsley. *RODBC: ODBC Database Access*, 2013. R package version 1.3-10.

[135] B. D. Ripley. Using databases with R. *R News*, 1(1):18–20, 2001.

[136] M. L. Rizzo. *Statistical Computing with R*. CRC Press, Boca Raton, FL, 2007.

[137] J. P. Romano and A. F. Siegel. *Counterexamples in Probability and Statistics*. Duxbury Press, 1986.

[138] P. R. Rosenbaum and D. B. Rubin. Reducing bias in observational studies using subclassification on the propensity score. *Journal of the American Statistical Association*, 79:516–524, 1984.

[139] P. R. Rosenbaum and D. B. Rubin. Constructing a control group using multivariate matched sampling methods that incorporate the propensity score. *The American Statistician*, 39:33–38, 1985.

[140] RStudio. *ggvis: Interactive Grammar of Graphics*, 2014. R package version 0.4.

[141] RStudio. *shiny: Web Application Framework for R*, 2014. R package version 0.10.2.1.

[142] D. B. Rubin. Multiple imputation after 18+ years. *Journal of the American Statistical Association*, 91:473–489, 1996.

[143] R. Saitz, N. J. Horton, M. J. Larson, M. Winter, and J. H. Samet. Primary medical care and reductions in addiction severity: a prospective cohort study. *Addiction*, 100(1):70–78, 2005.

[144] R. Saitz, M. J. Larson, N. J. Horton, M. Winter, and J. H. Samet. Linkage with primary medical care in a prospective cohort of adults with addictions in inpatient detoxification: room for improvement. *Health Services Research*, 39(3):587–606, 2004.

[145] J. H. Samet, M. J. Larson, N. J. Horton, K. Doyle, M. Winter, and R. Saitz. Linking alcohol and drug dependent adults to primary medical care: a randomized controlled trial of a multidisciplinary health intervention in a detoxification unit. *Addiction*, 98(4):509–516, 2003.

[146] J.-M. Sarabia, E. Castillo, and D. J. Slottje. An ordered family of Lorenz curves. *Journal of Econometrics*, 91:43–60, 1999.

[147] D. Sarkar. *Lattice: Multivariate Data Visualization with R*. Springer, New York, 2008.

[148] C.-E. Särndal, B. Swensson, and J. Wretman. *Model Assisted Survey Sampling*. Springer-Verlag, New York, 1992.

[149] J. L. Schafer. *Analysis of Incomplete Multivariate Data*. Chapman & Hall, London, 1997.

[150] J. L. Schafer. *mix: Estimation/Multiple Imputation for Mixed Categorical and Continuous Data*, 2010. R package version 1.0-8.

[151] M. E. Schaffer. *rtf: Rich Text Format Output*, 2013. R package version 0.4-11.

[152] B. Schloerke, J. Crowley, D. Cook, H. Hofmann, H. Wickham, F. Briatte, and M. Marbach. *GGally: Extension to ggplot2*, 2014. R package version 0.4.8.

[153] M. Schwartz. *WriteXLS: Cross-platform Perl Based R function to Create Excel 2003 (XLS) and Excel 2007 (XLSX) Files*, 2014. R package version 3.5.1.

[154] R. L. Schwartz, b. d. foy, and T. Phoenix. *Learning Perl (sixth edition)*. O'Reilly and Associates, 2011.

[155] L. Scrucca. *dispmod: Dispersion Models*, 2012. R package version 1.1.

[156] G. A. F. Seber and C. J. Wild. *Nonlinear Regression*. John Wiley & Sons, Hoboken, NJ, 1989.

[157] J. S. Sekhon. Multivariate and propensity score matching software with automated balance optimization: the Matching package for R. *Journal of Statistical Software*, 42(7):1–52, 2011.

[158] C. W. Shanahan, A. Lincoln, N. J. Horton, R. Saitz, M. J. Larson, and J. H. Samet. Relationship of depressive symptoms and mental health functioning to repeat detoxification. *Journal of Substance Abuse Treatment*, 29:117–123, 2005.

[159] M. S. Shotwell. *sas7bdat: SAS Database Reader*, 2014. R package version 0.5.

[160] T. Sing, O. Sander, N. Beerenwinkel, and T. Lengauer. ROCR: visualizing classifier performance in R. *Bioinformatics*, 21(20):3940–3941, 2005.

[161] T. Sing, O. Sander, N. Beerenwinkel, and T. Lengauer. ROCR: visualizing classifier performance in R. *Bioinformatics*, 21(20): 2005.

[162] S. Sturtz, U. Ligges, and A. Gelman. R2WinBUGS: A package for running WinBUGS from R. *Journal of Statistical Software*, 12(3):1–16, 2005.

[163] Y.-S. Su and M. Yajima. *R2jags: A Package for Running JAGS from R*, 2014. R package version 0.04-03.

[164] B. G. Tabachnick and L. S. Fidell. *Using Multivariate Statistics (fifth edition)*. Allyn & Bacon, Boston, 2007.

[165] S. M. M. Tahaghoghi and H. E. Williams. *Learning MySQL*. O'Reilly Media, Sebastopol, CA, 2006.

[166] T. M. Therneau and P. M. Grambsch. *Modeling Survival Data: Extending the Cox Model*. Springer, New York, 2000.

[167] T.M. Therneau, B. Atkinson, and B. Ripley. *rpart: Recursive Partitioning*, 2014. R package version 4.1-8.

[168] A. Thomas, B. O'Hara, U. Ligges, and S. Sturtz. Making BUGS open. *R News*, 6(1):12–17, 2006.

[169] R. Tibshirani. Regression shrinkage and selection via the lasso. *Journal of the Royal Statistical Society B*, 58(1), 1996.

[170] E. R. Tufte. *Envisioning Information*. Graphics Press, Cheshire, CT, 1990.

[171] E. R. Tufte. *Visual Explanations: Images and Quantities, Evidence and Narrative*. Graphics Press, Cheshire, CT, 1997.

[172] E. R. Tufte. *Visual Display of Quantitative Information (second edition)*. Graphics Press, Cheshire, CT, 2001.

[173] E. R. Tufte. *Beautiful Evidence*. Graphics Press, Cheshire, CT, 2006.

[174] J. W. Tukey. *Exploratory Data Analysis*. Addison Wesley, 1977.

[175] K. Ushey, J. McPherson, J. Cheng, and J. J. Allaire. *packrat: A Dependency Management System for Projects and Their R Package Dependencies*, 2014. R package version 0.4.1-1.

[176] S. van Buuren. *Flexible Imputation of Missing Data*. CRC Press, Boca Raton, FL, 2012.

[177] S. van Buuren, H. C. Boshuizen, and D. L. Knook. Multiple imputation of missing blood pressure covariates in survival analysis. *Statistics in Medicine*, 18:681–694, 1999.

[178] S. van Buuren and K. Groothuis-Oudshoorn. mice: Multivariate imputation by chained equations in R. *Journal of Statistical Software*, 45(3):1–67, 2011.

[179] W. N. Venables and B. D. Ripley. *Modern Applied Statistics with S (fourth edition)*. Springer, New York, 2002.

[180] W. N. Venables, D. M. Smith, and the R Core Team. An introduction to R: notes on R: a programming environment for data analysis and graphics, version 3.0.2. http://cran.r-project.org/doc/manuals/R-intro.pdf, accessed October 27, 2013, 2013.

[181] J. Verzani. *Using R for Introductory Statistics*. CRC Press, Boca Raton, FL, 2005.

[182] G. R. Warnes. *gmodels: Various R Programming Tools for Model Fitting*, 2013. R package version 2.15.4.1.

[183] G. R. Warnes, B. Bolker, G. Gorjanc, G. Grothendieck, A. Korosec, T. Lumley, D. MacQueen, A. Magnusson, and J. Rogers. *gdata: Various R Programming Tools for Data Manipulation*, 2014. R package version 2.13.3.

[184] G. R. Warnes, B. Bolker, and T. Lumley. *gtools: Various R Programming Tools*, 2014. R package version 3.4.1.

[185] B. West, K. B. Welch, and A. T. Galecki. *Linear Mixed Models: A Practical Guide Using Statistical Software*. CRC Press, Boca Raton, FL, 2006.

[186] I. R. White and P. Royston. Imputing missing covariate values for the Cox model. *Statistics in Medicine*, 28:1982–1998, 2009.

[187] H. Wickham. Reshaping data with the reshape package. *Journal of Statistical Software*, 21(12), 2007.

[188] H. Wickham. *ggplot2: Elegant Graphics for Data Analysis*. Springer, New York, 2009.

[189] H. Wickham. ASA 2009 data expo. *Journal of Computational and Graphical Statistics*, 20(2):281–283, 2011.

[190] H. Wickham. The Split-Apply-Combine strategy for data analysis. *Journal of Statistical Software*, 40(1):1–29, 2011.

[191] H. Wickham. *httr: Tools for working with URLs and HTTP*, 2014. R package version 0.5.

[192] H. Wickham. *tidyr: Easily Tidy Data with Spread and Gather Functions*, 2014. R package version 0.1.

[193] H. Wickham and W. Chang. *devtools: Tools to Make Developing R Code Easier*, 2014. R package version 1.6.

[194] H. Wickham and R. Francois. *dplyr: A Grammar of Data Manipulation*, 2014. R package version 0.3.

[195] S. Wilhelm and B. G. Manjunath. *tmvtnorm: Truncated Multivariate Normal and Student t Distribution*, 2014. R package version 1.4-9.

[196] L. Wilkinson. Dot plots. *The American Statistician*, 53(3):276–281, 1999.

[197] L. Wilkinson, D. Wills, D. Rope, A. Norton, and R. Dubbs. *The Grammar of Graphics (second edition)*. Springer-Verlag, New York, 2005.

[198] J. D. Wines, R. Saitz, N. J. Horton, C. Lloyd-Travaglini, and J. H. Samet. Overdose after detoxification: a prospective study. *Drug and Alcohol Dependence*, 89:161–169, 2007.

[199] Y. Xie. *Dynamic Documents with R and knitr*. CRC Press, Boca Raton, FL, 2014.

[200] Y. Xie. *knitr: A General-Purpose Package for Dynamic Report Generation in R*, 2014. R package version 1.6.

[201] T. W. Yee. The VGAM package for categorical data analysis. *Journal of Statistical Software*, 32(10):1–34, 2010.

[202] D. Zamar, B. McNeney, and J. Graham. elrm: Software implementing exact-like inference for logistic regression models. *Journal of Statistical Software*, 21(3), 2007.

[203] A. Zeileis and T. Hothorn. Diagnostic checking in regression relationships. *R News*, 2(3):7–10, 2002.

Appendix D

Indices

Separate indices are provided for subject (concept or task) and R command. References to the examples are denoted in *italics*.

D.1 Subject index

3-D
    histogram, 128
    plot, 130
95% confidence interval
    mean, 52
    proportion, 53

absolute value, 36
accelerated failure time model, 99
access
    Dropbox files, 6
    elements in R, 221
    files, 50
    variables, 11
add
    lines to plot, 146
    marginal rug plot, 147
    matrices, 39
    noise, 146
    normal density, 147
    straight line, 145
    text, 147
    variables, 13
age variable, *64*, 239
agreement, 54
AIC, *86*, 102
airline delays, *207*
Akaike information criterion (AIC), *86*, 102
alcohol abuse, 241

alcoholic drinks
    HELP dataset, 240
Allaire, J.J., xxii
altitude, *193*
Amazon sales rank, *195*
analysis of variance
    interaction plot, 130
    one-way, 70
    two-way, 70, *84*
analytic power calculations, 58
and operator, *28*
angular plot, 131
annotating datasets, *26*
ANOVA
    interaction plot, 130
    one-way, 70
    tables, 102
Aotearoa (New Zealand), 211
API (application programming interface), *199, 200, 202*
Apple R FAQ, 213
application programming interface (API), *199, 200, 202*
arbitrary quantiles, 52
area under the curve, 132
ARIMA model, 98
arrays, *27*, 46
    extract elements, 223
arrows, 148

D.2 R index